Biology Modular Workbook Series

Skills in Biology

The Biozone Writing Team

Tracey Greenwood

Lissa Bainbridge-Smith

Richard Allan

Dan Butler

Published by:
Biozone International Ltd
109 Cambridge Road, Hamilton 3216, New Zealand

Second printing
Printed by FUSION PRINT GROUP LTD
using paper produced from renewable and waste materials

Distribution Offices:

United Kingdom & Europe	**Biozone Learning Media (UK) Ltd**, UK
	Telephone: +44 1283-553-257
	Fax: +44 1283-553-258
	Email: sales@biozone.co.uk
	Website: www.biozone.co.uk
USA, Canada, Asia, South America, Africa	**Biozone International Ltd**, New Zealand
	Telephone: +64 7-856-8104
	Toll FREE: 1866-556-2710 (USA-Canada only)
	Freefax: 1-800-717-8751 (USA-Canada only)
	Fax: +64 7-856-9243
	Email: sales@biozone.co.nz
	Website: www.biozone.co.nz
Australia	**Biozone Learning Media Australia**, Australia
	Telephone: +61 7-5535-4896
	Fax: +61 7-5508-2432
	Email: sales@biozone.com.au
	Website: www.biozone.com.au

© 2007 **Biozone International Ltd**
ISBN: 978-1-877462-14-6

First Edition 2006
Second Edition 2007

Front cover photographs:
Girl with microscope. Image ©2005 JupiterImages Corporation www.clipart.com
Using data loggers in the field. Image courtesy of Pasco ©2004

NOTICE TO TEACHERS

No part of this workbook may be photocopied under any circumstances. This is a legal precondition of sale and includes a ban on photocopying under any photocopying licence scheme
(such as the Copyright Licensing Agency).

Biology Modular Workbook Series

The Biozone *Biology Modular Workbook Series* has been developed to meet the demands of customers with the requirement for a modular resource which can be used in a flexible way. Like Biozone's popular Student Resource and Activity Manuals, these workbooks provide a collection of visually interesting and accessible activities, which cater for students with a wide range of abilities and background. The workbooks are divided into a series of chapters, each comprising an introductory section with detailed learning objectives and useful resources, and a series of write-on activities ranging from paper practicals and data handling exercises, to questions requiring short essay style answers. A new feature in this edition is the inclusion of page tabs identifying "**Related activities**" in the workbook. These will help students to locate related material for help or additional detail if it is required. Material for these workbooks has been drawn from Biozone's popular, widely used manuals, but the workbooks have been structured with ease of use and flexibility in mind. During the development of this series, we have taken the opportunity to improve the design and content, while retaining the basic philosophy of a student-friendly resource which spans the gulf between textbook and study guide. With its unique, highly visual presentation, it is possible to engage and challenge students, increase their motivation and empower them to take control of their learning.

Skills in Biology

This title in the *Biology Modular Workbook Series* provides students with guidelines for planning and executing biological investigations in both the laboratory and the field. It comprises four chapters, which correspond to four areas relating to student skills: the design and analysis of experimental work, practical skills in field work, microscopy and microbiological techniques, and biological classification. These areas are explained through a series of one, two, or three page activities, each of which explores a specific topic area (e.g. data transformation). Model answers (on CD-ROM) accompany each order free of charge. Biozone's supplementary resource, the Teacher Resource CD-ROM (TRC) is also available for purchase and provides working spreadsheets for the statistical activities in this workbook. *Skills in Biology* is a student-centered resource. Students completing the activities, in concert with their other classroom and practical work, will consolidate existing knowledge and develop and practise the skills that they will use throughout their courses in biology. This workbook may be used in the classroom or at home as a supplement to a standard textbook. Some activities are introductory in nature, while others may be used to consolidate and test concepts already covered by other means (e.g. microscopy). Biozone has a commitment to produce a cost-effective, high quality resource which acts as a student's companion throughout their biology study. Please do not photocopy from this workbook; we cannot afford to provide single copies of workbooks to schools and continue to develop, update, and improve the material they contain.

Acknowledgements and Photo Credits

Royalty free images, purchased by Biozone International Ltd, are used throughout this workbook and have been obtained from the following sources: Corel Corporation from various titles in their Professional Photos CD-ROM collection; IMSI (International Microcomputer Software Inc.) images from IMSI's MasterClips® and MasterPhotos™ Collection, 1895 Francisco Blvd. East, San Rafael, CA 94901-5506, USA; ©1996 Digital Stock, Medicine and Health Care collection; ©Hemera Technologies Inc., 1997-2001; © 2005 JupiterImages Corporation www.clipart.com; ©Click Art, ©T/Maker Company; ©1994., ©Digital Vision; Gazelle Technologies Inc.; PhotoDisc®, Inc. USA, www.photodisc.com. The authors would also like to thank those who have contributed towards this edition: • Sam Banks for his photograph of wombat scat • Pasco for their use of images of sampling using probeware • Vernier for use of their image of a respiration chamber • Dave Ward, Sirtrack Ltd, for photographs and information on radio-tracking • Campus Photography, University of Waikato for photographs of equipment used for monitoring physical factors. Photos kindly provided by individuals or corporations have been indentified by way of coded credits as follows: **BOB**: Barry O'Brien (Uni. of Waikato), **BF**: Brian Finerran (University of Canterbury), **BH**: Brendan Hicks (University of Waikato), **CDC**: Centers for Disease Control and Prevention, Atlanta, USA, **COD**: Colin O'Donnell (Dept of Conservation, NZ), **GU**: Graeme Ussher (University of Auckland), **EII**: Education Interactive Imaging, **EW**: Environment Waikato, **FRI**: Forest Research Institute, **GW**: Graham Walker, **HF**: Halema Flannagan, **JDG**: John Green (University of Waikato), **KL**-Sirtrack: Kevin Lay (Sirtrack Ltd), **PASCO**: Pasco Probeware, **PH**: Phil Herrity, **RA**: Richard Allan, **RCN**: Ralph Cocklin, **Sirtrack**: Sirtrack Ltd, **VM**: Villa Maria Wines, **VU**: Victoria University, NZ, **WMU**: Waikato Microscope Unit.

Also in this series:

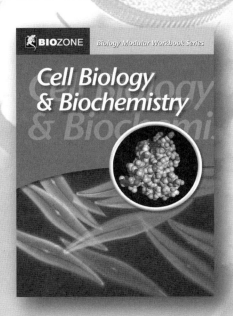

Cell Biology & Biochemistry
ISBN: 1-877329-75-4

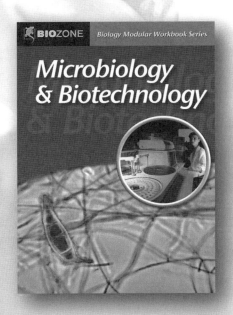

Microbiology & Biotechnology
ISBN: 978-1-877462-12-2

Contents

Note to the Teacher and Acknowledgements iii
How to Use This Manual 1
Activity Pages..2
Explanation of Terms..3
Using the Internet 4
Concept Map for Skills in Biology..........................6
Resources Information 7

Biological Investigations

Objectives and Resources 8
Terms and Notation 10
The Scientific Method..11
A Guide to Research Projects 13
Choosing Your Topic 14
Hypotheses and Predictions 15
Planning an Investigation 17
Designing Your Experiment 19
Recording Results 21
Tables and Graphs 22
Transforming Raw Data 23
Types of Graphs 25
Drawing Bar Graphs 27
Drawing Histograms 28
Drawing Pie Graphs 29
Drawing Kite Graphs 30
Drawing Scatter Plots 31
Drawing Line Graphs 32
Interpreting Line Graphs 35
Taking the Next Step 37
Descriptive Statistics 39
The Reliability of the Mean 41
Linear Regression 43
Non-Linear Regression 45
The Student's *t* Test 46
Student's *t* Test Exercise 47
Comparing More Than Two Groups 49
Analysis of Variance 50
Using the Chi-Squared Test in Ecology 52
Chi-Squared Exercise in Ecology 53
Using the Chi-Squared Test in Genetics 54
Chi-Squared Exercise in Genetics 55
△ The Structure of a Report 56
Writing the Methods 57
Writing Your Results 58
Writing Your Discussion 59
Report Checklist 60
Citing and Listing References 61

Field Studies

Objectives and Resources 63
Sampling Populations 64
Designing Your Field Study 65
Monitoring Physical Factors 67
Quadrat Sampling 69
Quadrat-Based Estimates 70
Sampling a Leaf Litter Population 71
Transect Sampling 73
Sampling Animal Populations 75
Mark and Recapture Sampling 77
Indirect Sampling 79
Sampling Using Radio-tracking 81

Classification of Organisms

Objectives and Resources 83
The New Tree of Life 84
New Classification Schemes.............................. 85
Features of Taxonomic Groups 86
Classification System 91
Classification Keys 93
☆ Keying Out Plant Species................................... 95
Features of the Five Kingdoms 96
The Classification of Life 97
Features of Animal Taxa................................... 103
Features of Fungi and Plants 105

Laboratory Techniques

Objectives and Resources 106
Biological Drawings 107
△ Optical Microscopes 109
Electron Microscopes 111
Interpreting Electron Micrographs 113
Identifying Cell Structures 115
Biochemical Tests 116
Differential Centrifugation 117
Gel Elecrophoresis 118
Analyzing a DNA Sample 119
Techniques in Microbial Culture 120
Strain Isolation 121
Serial Dilution 122
Plant Tissue Culture 123

INDEX 124

CODES: △ **Upgraded** this edition   ☆ **New** this edition   **Activity** is marked: to be done; ▪   when completed ✓

How to Use this Workbook

Skills in Biology is designed to provide biology students with a resource that will make the acquisition of investigative skills easier and more enjoyable. The ability to plan investigative studies, present and analyse data, and accurately report your findings are core skills in most biology curricula. This workbook is suitable for all students of biology, and will reinforce and extend the ideas introduced and developed by your teacher. It is **not a textbook**; its aim is to complement the texts written for your particular course. *Skills in Biology* provides the following useful resources in each chapter:

Guidance Provided for Each Topic

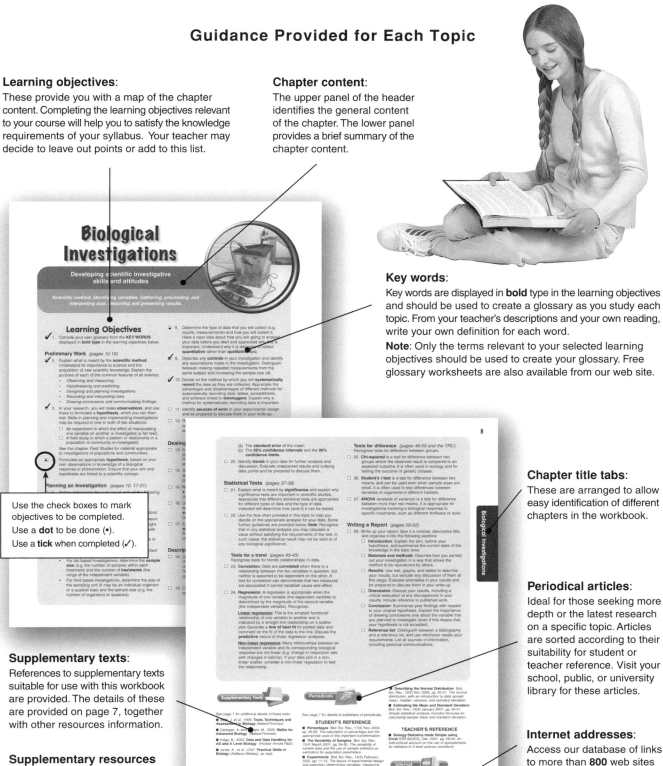

Learning objectives:
These provide you with a map of the chapter content. Completing the learning objectives relevant to your course will help you to satisfy the knowledge requirements of your syllabus. Your teacher may decide to leave out points or add to this list.

Chapter content:
The upper panel of the header identifies the general content of the chapter. The lower panel provides a brief summary of the chapter content.

Key words:
Key words are displayed in **bold** type in the learning objectives and should be used to create a glossary as you study each topic. From your teacher's descriptions and your own reading, write your own definition for each word.

Note: Only the terms relevant to your selected learning objectives should be used to create your glossary. Free glossary worksheets are also available from our web site.

Use the check boxes to mark objectives to be completed.
Use a **dot** to be done (•).
Use a **tick** when completed (✓).

Chapter title tabs:
These are arranged to allow easy identification of different chapters in the workbook.

Periodical articles:
Ideal for those seeking more depth or the latest research on a specific topic. Articles are sorted according to their suitability for student or teacher reference. Visit your school, public, or university library for these articles.

Supplementary texts:
References to supplementary texts suitable for use with this workbook are provided. The details of these are provided on page 7, together with other resources information.

Supplementary resources
Biozone's Presentation MEDIA are noted where appropriate. Supporting spreadsheet activities, and computer software and videos relevant to every topic in the manual are provided on Biozone's **Teacher Resource CD-ROM**.

Internet addresses:
Access our database of links to more than **800** web sites (updated regularly) relevant to the topics covered. Go to Biozone's own web site: **www.thebiozone.com** and link to listed sites using the *BioLinks* button.

Activity Pages

The activities and exercises make up most of the content of this workbook. They are designed to reinforce the concepts you have learned about in the topic. Your teacher may use the activity pages to introduce a topic for the first time, or you may use them to revise ideas already covered. They are excellent for use in the classroom, and as homework exercises and revision. In most cases, the activities should not be attempted until you have carried out the necessary background reading from your textbook. Model answers providing suggested answers for each activity are available at the end of the book.

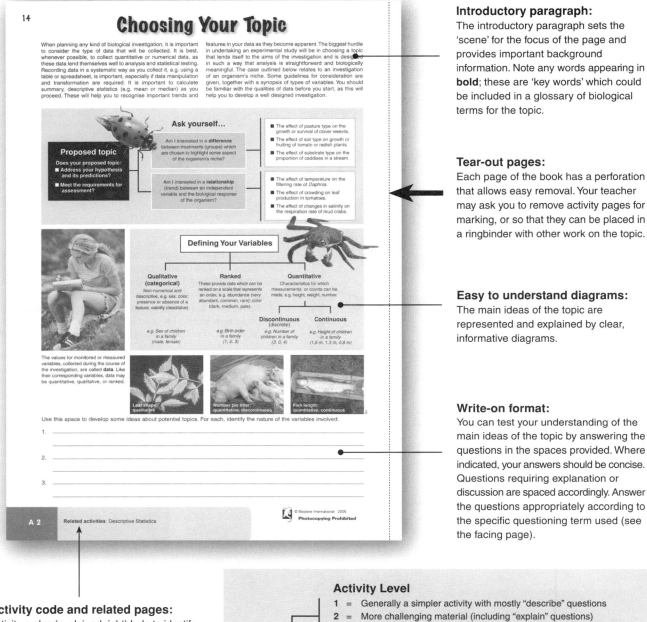

Introductory paragraph:
The introductory paragraph sets the 'scene' for the focus of the page and provides important background information. Note any words appearing in **bold**; these are 'key words' which could be included in a glossary of biological terms for the topic.

Tear-out pages:
Each page of the book has a perforation that allows easy removal. Your teacher may ask you to remove activity pages for marking, or so that they can be placed in a ringbinder with other work on the topic.

Easy to understand diagrams:
The main ideas of the topic are represented and explained by clear, informative diagrams.

Write-on format:
You can test your understanding of the main ideas of the topic by answering the questions in the spaces provided. Where indicated, your answers should be concise. Questions requiring explanation or discussion are spaced accordingly. Answer the questions appropriately according to the specific questioning term used (see the facing page).

Activity code and related pages:
Activity codes (explained right) help to identify the type of activities and the skills they require. Most activities require knowledge recall as well as the application of knowledge to explain observations or predict outcomes.

Use the **Related activities** indicated to visit pages that may help you with understanding the material or answering the questions.

Activity Level
1 = Generally a simpler activity with mostly "describe" questions
2 = More challenging material (including "explain" questions)
3 = Challenging content or questions (more "discuss" questions)

Type of Activity
D = Includes some data handling and/or interpretation
P = includes a paper practical
R = May require research outside the page
A = Includes application of knowledge to solve a problem
E = Extension material

Explanation of Terms

Questions come in a variety of forms. Whether you are studying for an exam or writing an essay, it is important to understand exactly what the question is asking. A question has two parts to it: one part of the question will provide you with information, the second part of the question will provide you with instructions as to how to answer the question. Following these instructions is most important. Often students in examinations know the material but fail to follow instructions and do not answer the question appropriately. Examiners often use certain key words to introduce questions. Look out for them and be clear as to what they mean. Below is a description of terms commonly used when asking questions in biology.

Commonly used Terms in Biology

The following terms are frequently used when asking questions in examinations and assessments. Students should have a clear understanding of each of the following terms and use this understanding to answer questions appropriately.

Account for: Provide a satisfactory explanation or reason for an observation.

Analyze: Interpret data to reach stated conclusions.

Annotate: Add **brief** notes to a diagram, drawing or graph.

Apply: Use an idea, equation, principle, theory, or law in a new situation.

Appreciate: To understand the meaning or relevance of a particular situation.

Calculate: Find an answer using mathematical methods. Show the working unless instructed not to.

Compare: Give an account of similarities and differences between two or more items, referring to both (or all) of them throughout. Comparisons can be given using a table. Comparisons generally ask for similarities more than differences (see contrast).

Construct: Represent or develop in graphical form.

Contrast: Show differences. Set in opposition.

Deduce: Reach a conclusion from information given.

Define: Give the precise meaning of a word or phrase as concisely as possible.

Derive: Manipulate a mathematical equation to give a new equation or result.

Describe: Give a detailed account, including all the relevant information.

Design: Produce a plan, object, simulation or model.

Determine: Find the only possible answer.

Discuss: Give an account including, where possible, a range of arguments, assessments of the relative importance of various factors, or comparison of alternative hypotheses.

Distinguish: Give the difference(s) between two or more different items.

Draw: Represent by means of pencil lines. Add labels unless told not to do so.

Estimate: Find an approximate value for an unknown quantity, based on the information provided and application of scientific knowledge.

Evaluate: Assess the implications and limitations.

Explain: Give a clear account including causes, reasons, or mechanisms.

Identify: Find an answer from a number of possibilities.

Illustrate: Give concrete examples. Explain clearly by using comparisons or examples.

Interpret: Comment upon, give examples, describe relationships. Describe, then evaluate.

List: Give a sequence of names or other brief answers with no elaboration. Each one should be clearly distinguishable from the others.

Measure: Find a value for a quantity.

Outline: Give a brief account or summary. Include essential information only.

Predict: Give an expected result.

Solve: Obtain an answer using algebraic and/or numerical methods.

State: Give a specific name, value, or other answer. No supporting argument or calculation is necessary.

Suggest: Propose a hypothesis or other possible explanation.

Summarize: Give a brief, condensed account. Include conclusions and avoid unnecessary details.

In Conclusion

Students should familiarise themselves with this list of terms and, where necessary throughout the course, they should refer back to them when answering questions. The list of terms mentioned above is not exhaustive and students should compare this list with past examination papers / essays etc. and add any new terms (and their meaning) to the list above. The aim is to become familiar with interpreting the question and answering it appropriately.

Using the Internet

The internet is a powerful resource for locating information. There are several key areas of Biozone's web site that may be of interest to you. Go to the **BioLinks** area to browse through the hundreds of web sites hosted by other organizations. These sites provide a supplement to the activities provided in our workbooks and have been selected on the basis of their accurate, current, and relevant content. We have also provided links to biology-related **podcasts** and **RSS newsfeeds**. These provide regularly updated information about new discoveries in biology; perfect for those wanting to keep abreast of changes in this dynamic field.

The BIOZONE website: www.thebiozone.com

The current internet address (URL) for the web site is displayed here. You can type a new address directly into this space.

Use Google to search for web sites of interest. The more precise your search words are, the better the list of results. EXAMPLE: If you type in "biotechnology", your search will return an overwhelmingly large number of sites, many of which will not be useful to you. Be more specific, e.g. "biotechnology medicine DNA uses".

Find out about our superb **Presentation Media**. These slide shows are designed to provide in-depth, highly accessible illustrative material and notes on specific areas of biology.

Podcasts: Access the latest news as audio files (mp3) that may be downloaded to your ipod (mp3 player) or played directly off your computer.

RSS Newsfeeds: See breaking news and major new discoveries in biology directly from our web site.

Access the **BioLinks** database of web sites related to each major area of biology.

The **Resource Hub** provides links to the supporting resources referenced in the workbook. These resources include comprehensive and supplementary texts, biology dictionaries, computer software, videos, and science supplies.

News: Find out about product announcements, shipping dates, and workshops and trade displays by Biozone at teachers' conferences around the world.

Photocopying Prohibited © Biozone International 2006-2007

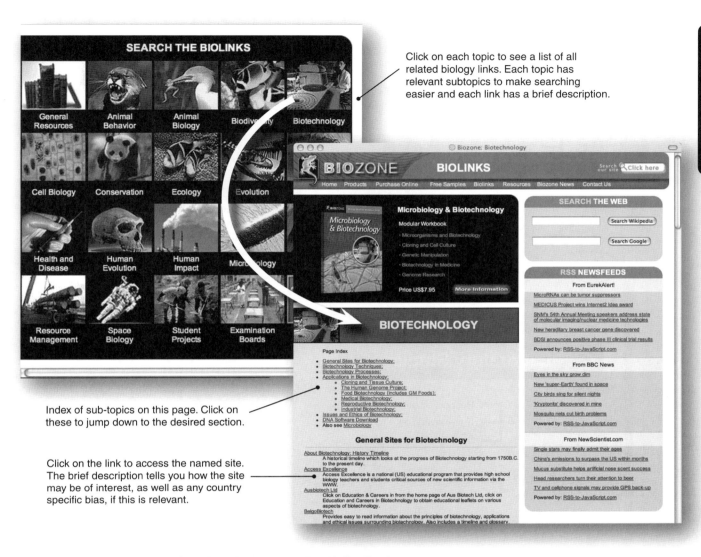

Click on each topic to see a list of all related biology links. Each topic has relevant subtopics to make searching easier and each link has a brief description.

Index of sub-topics on this page. Click on these to jump down to the desired section.

Click on the link to access the named site. The brief description tells you how the site may be of interest, as well as any country specific bias, if this is relevant.

Access free samples from Biozone's product range. Samples include:

- Activities from the Student Workbooks
- Samples of the Presentation Media
- Sample crossword
- Glossary worksheets
- Choice flowchart for data analysis

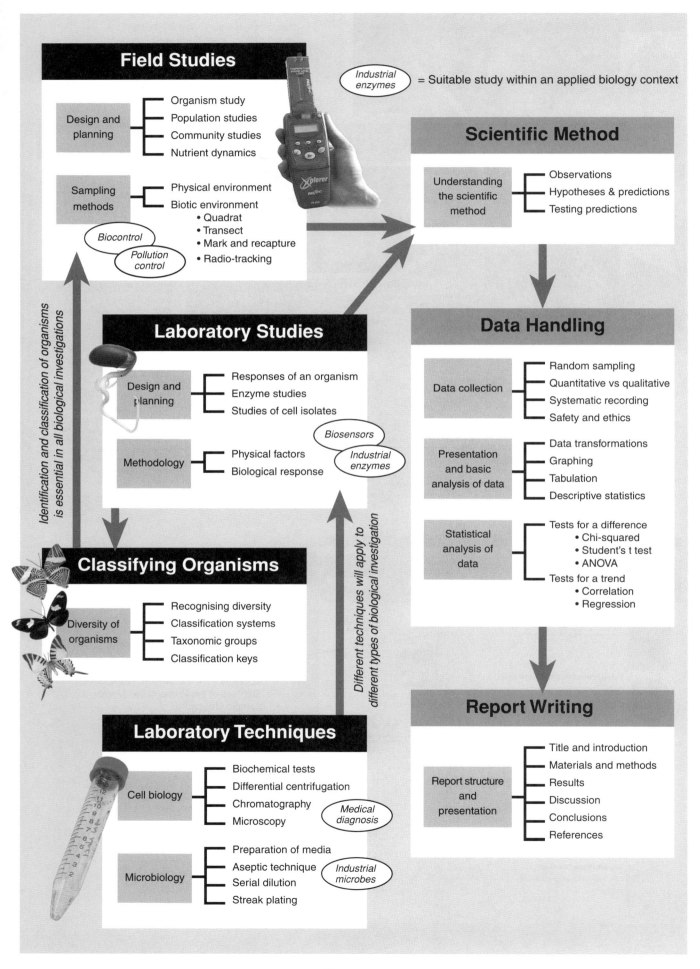

Resources Information

Your set textbook should always be a starting point for information, but there are also many other resources available. A list of some readily available resources is provided below. Access to the publishers of these resources can be made directly from Biozone's web site through our resources hub: www.thebiozone.com/resource-hub.html. Please note that listing any product in this workbook does not, in any way, denote Biozone's endorsement of that product and Biozone does not have any business affiliation with the publishers listed herein.

Supplementary Texts

Adds, J., E. Larkcom, R. Miller, & R. Sutton, 1999. **Tools, Techniques and Assessment in Biology**, 160 pp. **ISBN**: 0-17-448273-6
A course guide covering basic lab protocols, microscopy, quantitative lab and field techniques, advanced DNA techniques and tissue culture, data handling, and exam prep.

Cadogan, A. and Ingram, M., 2002
Maths for Advanced Biology
Publisher: NelsonThornes
ISBN: 0-7487-6506-9
Comments: *Covers the maths requirements of senior level biology. Includes worked examples.*

Indge, B., 2003
Data and Data Handling for AS and A Level Biology, 128 pp.
Publisher: Hodder Arnold H&S
ISBN: 1340856475
Comments: *Examples and practice exercises to improve skills in data interpretation and analysis.*

Jones, A., R. Reed, & J. Weyers, 4th ed. 2007
Practical Skills in Biology, approx. 300 pp.
Publisher: Pearson
ISBN: 978-0-131775-09-3
Comments: *Provides information on all aspects of experimental and field design, implementation, and data analysis.*

Periodicals, Magazines, and Journals

Biological Sciences Review: *An informative quarterly publication for biology students.*
Enquiries:
Philip Allan Publishers, Market Place,
Deddington, Oxfordshire OX 15 OSE
Tel: 01869 338652
Fax: 01869 338803
E-mail: sales@philipallan.co.uk

New Scientist: *Widely available weekly magazine with research summaries and features.* Enquiries:
Reed Business Information Ltd, 51 Wardour St. London WIV 4BN **Tel**: (UK and intl):+44 (0) 1444 475636 **E-mail**: ns.subs@qss-uk.com or subscribe from their web site.

Scientific American: *A monthly magazine containing specialist features. Articles range in level of reading difficulty and assumed knowledge.* Subscription enquiries: 415 Madison Ave. New York. NY10017-1111 **Tel**: (outside North America): 515-247-7631 **Tel**: (US & Canada): 800-333-1199

The American Biology Teacher: *The peer-reviewed journal of the NABT. Published nine times a year and containing information and activities relevant to biology teachers.* Contact: NABT, 12030 Sunrise Valley Drive, #110, Reston, VA 20191-3409
Web: www.nabt.org

Biology Dictionaries

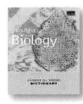

Clamp, A.
AS/A-Level Biology. Essential Word Dictionary, 2000, 161 pp. Philip Allan Updates.
ISBN: 0-86003-372-4.
Essential words for AS and A2. Concise definitions are supported by further explanation and illustrations where required.

Hale, W.G. **Collins: Dictionary of Biology** 4 ed. 2005, 528 pp. Collins.
ISBN: 0-00-720734-4.
Updated to take in the latest developments in biology and now internet-linked. (§ This latest edition is currently available only in the UK. The earlier edition, ISBN: 0-00-714709-0, is available though amazon.com in North America).

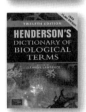

Henderson, I.F, W.D. Henderson, and E. Lawrence. **Henderson's Dictionary of Biological Terms**, 1999, 736 pp. Prentice Hall.
ISBN: 0582414989
An updated edition, rewritten for clarity, and reorganized for ease of use. An essential reference and the dictionary of choice for many.

Market House Books (compiled by).
Oxford Dictionary of Biology 5 ed., 2004, 698 pp. Oxford University Press.
ISBN: 0198609175. *Revised and updated, with many new entries. This edition contains biographical entries on key scientists and comprehensive coverage of terms in biology, biophysics, and biochemistry.*

McGraw-Hill (ed). **McGraw-Hill Dictionary of Bioscience**, 2 ed., 2002, 662 pp. McGraw-Hill.
ISBN: 0-07-141043-0
22 000 entries encompassing more than 20 areas of the life sciences. It includes synonyms, acronyms, abbreviations, and pronunciations for all terms. Accessible, yet comprehensive.

Rudin, N.
Dictionary of Modern Biology (1997), 504 pp. Barron's Educational Series Inc
ISBN: 0812095162.
More than 6000 terms in biosciences defined for college level students. Includes extensive cross referencing and several useful appendices.

Biological Investigations

Developing scientific investigative skills and attitudes

Scientific method. Identifying variables. Gathering, processing, and interpreting data. Reporting and presenting results.

Learning Objectives

☐ 1. Compile your own glossary from the **KEY WORDS** displayed in **bold type** in the learning objectives below.

Preliminary Work (pages 10-16)

☐ 2. Explain what is meant by the **scientific method**. Understand its importance to science and the acquisition of new scientific knowledge. Explain the purpose of each of the common features of all science:
- Observing and measuring.
- Hypothesising and predicting.
- Designing and planning investigations.
- Recording and interpreting data.
- Drawing conclusions and communicating findings.

☐ 3. In your research, you will make **observations**, and use these to formulate a **hypothesis**, which you can then test. Skills in planning and implementing investigations may be required in one or both of two situations:
  ☐ An experiment in which the effect of manipulating one variable on another is investigated (a fair test).
  ☐ A field study in which a pattern or relationship in a population or community is investigated.

See the chapter *Field Studies* for material appropriate to investigations of populations and communities.

☐ 4. Formulate an appropriate **hypothesis**, based on your own observations or knowledge of a biological response or phenomenon. Ensure that your aim and hypothesis are linked to a scientific concept.

Planning an Investigation (pages 10, 17-21)

☐ 5. Define and explain the purpose of each of the following variables in a controlled experiment:
- **Independent variable** (manipulated variable)
- **Dependent variable** (response variable)
- **Controlled variables** (to control nuisance factors)

☐ 6. For your own investigation distinguish clearly between:
- A **data value** for a particular **variable**, e.g. height.
- The individual sampling unit, e.g. a test-tube with an enzyme at a particular pH.
- The sample size, e.g. the number of test-tubes in each treatment.

☐ 7. Determine the amount of data that you need to collect in order to reasonably test your hypothesis.
- For lab based investigations, determine the **sample size** (e.g. the number of samples within each treatment) and the number of **treatments** (the range of the independent variable).
- For field based investigations, determine the size of the sampling unit (it may be an individual organism or a quadrat size) and the sample size (e.g. the number of organisms or quadrats).

☐ 8. Determine the type of data that you will collect (e.g. counts, measurements) and how you will collect it. Have a clear idea about how you are going to analyse your data before you start and appreciate why this is important. Understand why it is desirable to collect **quantitative** rather than **qualitative** data.

☐ 9. Describe any **controls** in your investigation and identify any assumptions made in the investigation. Distinguish between making repeated measurements from the same subject and increasing the sample size (**n**).

☐ 10. Decide on the method by which you will **systematically record** the data as they are collected. Appreciate the advantages and disadvantages of different methods for systematically recording data: tables, spreadsheets, and software linked to **dataloggers**. Explain why a method for systematically recording data is important.

☐ 11. Identify **sources of error** in your experimental design and be prepared to discuss them in your write-up.

☐ 12. Recognise that all biological investigations should be carried out with appropriate regard for safety and the well-being of living organisms and their environment.

Dealing with Data (pages 10, 21-36)

☐ 13. Collect and record data systematically according to your plan. Critically evaluate the **accuracy** of your methods for data collection, any **measurement errors**, and the repeatability (**precision**) of any measurements.

☐ 14. Demonstrate an ability to perform simple and appropriate **data transformations**, e.g. totals, percentages, increments, reciprocals, rates, and log.

☐ 15. Recognise the benefits of graphing data. Recognise the **x axis** and **y axis** of graphs and identify which variable (dependent or independent) is plotted on each.

☐ 16. Demonstrate an ability to plot data in an appropriate way using different methods of graphical presentation: **scatter plots**, **line graphs**, **pie graphs**, **bar graphs** (and column graphs), **kite graphs**, and **histograms**.

☐ 17. Explain what is meant by a '**line of best fit**' on a **scatter plot**. Understand that a line of best fit may be fitted by eye, or using a computer analysis. Draw 'lines of best fit' to scatter plots.

Descriptive Statistics (pages 39-42)

☐ 18. Distinguish between a **statistic** and a **parameter**. Demonstrate an understanding of the calculation and use of the following **descriptive statistics**:
  (a) Sample **mean** and **standard deviation**
  (b) **Median** and **mode** (calculated from your own, or second hand, data).

☐ 19. Calculate measures of dispersion for your data, related to the true population parameters. Consider:

- (a) The **standard error** of the mean.
- (b) The **95% confidence intervals** and the **95% confidence limits**.

☐ 20. Identify **trends** in your data for further analysis and discussion. Evaluate unexpected results and outlying data points and be prepared to discuss them.

Statistical Tests (pages 37-38)

☐ 21. Explain what is meant by **significance** and explain why significance tests are important in scientific studies. Appreciate that different statistical tests are appropriate for different types of data and the type of data collected will determine how (and if) it can be tested.

☐ 22. Use the flow chart provided in this topic to help you decide on the appropriate analysis for your data. Some further guidelines are provided below. **Note**: Recognise that in any statistical analysis you may calculate a value without satisfying the requirements of the test. In such cases, the statistical result may not be valid or of any biological significance.

Tests for a trend (pages 43-45)
Recognise tests for trends (relationships) in data.

☐ 23. **Correlation**: Data are **correlated** when there is a relationship between the two variables in question, but neither is assumed to be dependent on the other. A test for correlation can demonstrate that two measures are associated; it cannot establish cause and effect.

☐ 24. **Regression**: A regression is appropriate when the magnitude of one variable (the dependent variable) is determined by the magnitude of the second variable (the independent variable). Recognise:

Linear regression: This is the simplest functional relationship of one variable to another and is indicated by a straight line relationship on a scatter plot. Generate a **line of best fit** for plotted data and comment on the fit of the data to the line. Discuss the **predictive** nature of linear regression analyses.

Non-linear regression: Many relationships between an independent variable and its corresponding biological response are not linear (e.g. change in respiration rate with changes in salinity). If your data plot in a non-linear scatter, consider a non-linear regression to test the relationship.

Tests for difference (pages 46-55 and the TRC)
Recognise tests for difference between groups.

☐ 25. **Chi-squared** is a test for difference between two groups where the observed result is compared to an expected outcome. It is often used in ecology and for testing the outcome of genetic crosses.

☐ 26. **Student's *t* test** is a test for difference between two means, and can be used even when sample sizes are small. It is often used to test differences between densities of organisms in different habitats.

☐ 27. **ANOVA** (analysis of variance) is a test for difference between more than two means. It is appropriate for investigations involving a biological response to specific treatments, such as different fertilisers or soils.

Writing a Report (pages 56-62)

☐ 28. Write up your report. Give it a concise, descriptive title, and organise it into the following sections:
- ☐ **Introduction**: Explain the aim, outline your hypothesis, and summarise the current state of the knowledge in the topic area.
- ☐ **Materials and methods**: Describe how you carried out your investigation in a way that allows the method to be reproduced by others.
- ☐ **Results**: Use text, graphs, and tables to describe your results, but exclude any discussion of them at this stage. Evaluate anomalies in your results and be prepared to discuss them in your write-up.
- ☐ **Discussion**: Discuss your results, including a critical evaluation of any discrepancies in your results. Include reference to published work.
- ☐ **Conclusion**: Summarise your findings with respect to your original hypothesis. Explain the importance of drawing conclusions *only* about the variable that you planned to investigate (even if this means that your hypothesis is not accepted).
- ☐ **Reference list**: Distinguish between a bibliography and a reference list, and use whichever meets your requirements. List all sources of information, including personal communications.

See page 7 for additional details of these texts:

■ Adds, J. *et al.*, 1999. **Tools, Techniques and Assessment in Biology** (NelsonThornes).

■ Cadogan, A. and Ingram, M., 2002. **Maths for Advanced Biology** (NelsonThornes).

■ Indge, B., 2003. **Data and Data Handling for AS and A Level Biology** (Hodder Arnold H&S).

■ Jones, A., *et al.*, 2007. **Practical Skills in Biology** (Addison-Wesley), as required.

See page 7 for details of publishers of periodicals:

STUDENT'S REFERENCE

■ **Percentages** Biol. Sci. Rev., 17(2) Nov. 2004, pp. 28-29. *The calculation of percentage and the appropriate uses of this important transformation.*

■ **Drawing Graphs** Biol. Sci. Rev., 19(3) Feb. 2007, pp. 10-13. *A guide to creating graphs.*

■ **The Variability of Samples** Biol. Sci. Rev., 13(4) March 2001, pp. 34-35. *The variability of sample data and the use of sample statistics as estimators for population parameters.*

■ **Experiments** Biol. Sci. Rev., 14(3) February 2002, pp. 11-13. *The basics of experimental design and execution: determining variables, measuring them, and establishing a control.*

■ **Descriptive Statistics** Biol. Sci. Rev., 13 (5) May 2001, pp. 36-37. *A synopsis of descriptive statistics. The appropriate use of standard error and standard deviation is discussed.*

■ **Dealing with Data** Biol. Sci. Rev., 12 (4) March 2000, pp. 6-8. *A short account of the best ways in which to deal with the interpretation of graphically presented data in examinations.*

■ **Correlation** Biol. Sci. Rev., 14(3) February 2002, pp. 38-41. *An examination of the relationship between variables. An excellent synopsis.*

■ **Describing the Normal Distribution** Biol. Sci. Rev., 13(2) Nov. 2000, pp. 40-41. *The normal distribution, with an introduction to data spread, mean, median, variance, and standard deviation.*

■ **Estimating the Mean and Standard Deviation** Biol. Sci. Rev., 13(3) January 2001, pp. 40-41. *Simple statistical analysis. Includes formulae for calculating sample mean and standard deviation.*

TEACHER'S REFERENCE

■ **Biology Statistics made Simple using Excel** SSR 83(303), Dec. 2001, pp. 29-34. *An instructional account on the use of spreadsheets for statistics in A level science (excellent).*

See pages 4-5 for details of how to access **Bio Links** from our web site: **www.thebiozone.com** From Bio Links, access sites under the topics:

STUDENT PROJECTS: • A scientific report • Scientific investigation • Study skills - biology • The scientific method • Tree lupins ... *and others*

Working spreadsheets support this topic:

Teacher Resource CD-ROM
• Statistics spreadsheets

Terms and Notation

The definitions for some commonly encountered terms related to making biological investigations are provided below. Use these as you would use a biology dictionary when planning your investigation and writing up your report. It is important to be consistent with the use of terms i.e. use the same term for the same procedure or unit throughout your study. Be sure, when using a term with a specific statistical meaning, such as sample, that you are using the term correctly.

General Terms

Data: Facts collected for analysis.

Qualitative: Not quantitative. Described in words or terms rather than by numbers. Includes subjective descriptions in terms of variables such as color or shape.

Quantitative: Able to be expressed in numbers. Numerical values derived from counts or measurements.

The Design of Investigations

Hypothesis: A tentative explanation of an observation, capable of being tested by experimentation. Hypotheses are written as clear statements, not as questions.

Control treatment (control): A standard (reference) treatment that helps to ensure that responses to other treatments can be reliably interpreted. There may be more than one control in an investigation.

Dependent variable: A variable whose values are determined by another variable (the independent variable). In practice, the dependent variable is the variable representing the biological response.

Independent variable: A variable whose values are set, or systematically altered, by the investigator.

Controlled variables: Variables that may take on different values in different situations, but are controlled (fixed) as part of the design of the investigation.

Experiment: A contrived situation designed to test (one or more) hypotheses and their predictions. It is good practice to use sample sizes that are as large as possible for experiments.

Investigation: A very broad term applied to scientific studies; investigations may be controlled experiments or field based studies involving population sampling.

Parameter: A numerical value that describes a characteristic of a population (e.g. the mean height of all 18 year-old females).

Random sample: A method of choosing a sample from a population that avoids any subjective element. It is the equivalent to drawing numbers out of a hat, but using random number tables. For field based studies involving quadrats or transects, random numbers can be used to determine the positioning of the sampling unit.

Repeat / Trial: The entire investigation is carried out again at a different time. This ensures that the results are reproducible. Note that repeats or trials are not replicates in the true sense unless they are run at the same time.

Sample: A sub-set of a whole used to estimate the values that might have been obtained if every individual or response was measured. A sample is made up of **sampling units**, In lab based investigations, the sampling unit might be a test-tube, while in field based studies, the sampling unit might be an individual organism or a quadrat.

Sample size (n): The number of samples taken. In a field study, a typical sample size may involve 20-50 individuals or 20 quadrats. In a lab based investigation, a typical sample size may be two to three sampling units, e.g. two test-tubes held at 10°C.

Sampling unit: Sampling units make up the sample size. Examples of sampling units in different investigations are an individual organism, a test tube undergoing a particular treatment, an area (e.g. quadrat size), or a volume. The size of the sampling unit is an important consideration in most field studies where the area or volume of a habitat is being sampled.

Statistic: An estimate of a parameter obtained from a sample (e.g. the mean height of all 18 year-old females based on those in your class). *Compare this with the definition for parameter.*

Treatments: Well defined conditions applied to the sample units. The response of sample units to a treatment is intended to shed light on the hypothesis under investigation. What is often of most interest is the comparison of the responses to different treatments.

Variable: A factor in an experiment that is subject to change. Variables may be controlled (fixed), manipulated (systematically altered), or represent a biological response.

Precision and Significance

Accuracy: The correctness of the measurement (the closeness of the measured value to the true value). Accuracy is often a function of the calibration of the instrument used for measuring.

Measurement errors: When measuring or setting the value of a variable, there may be some difference between your answer and the 'right' answer. These errors are often as a result of poor technique or poorly set up equipment.

Objective measurement: Measurement not significantly involving subjective (or personal) judgment. If a second person repeats the measurement they should get the same answer.

Precision (of a measurement): The repeatability of the measurement. As there is usually no reason to suspect that a piece of equipment is giving inaccurate measures, making precise measurements is usually the most important consideration. You can assess or quantify the precision of any measurement system by taking repeated measurements from individual samples.

Precision (of a statistic): How close the statistic is to the value of the parameter being estimated. Also called **reliability**.

The Expression of Units

The value of a variable must be written with its units where possible. Common ways of recording measurements in biology are: volume in liters, mass in grams, length in meters, time in seconds. The following example shows different ways to express the same term. Note that ml and cm^3 are equivalent.

Oxygen consumption (milliliters per gram per hour)
Oxygen consumption ($mlg^{-1}h^{-1}$) or ($mLg^{-1}h^{-1}$)
Oxygen consumption (ml/g/h) or (mL/g/h)
Oxygen consumption/$cm^3g^{-1}h^{-1}$

Statistical significance: An assigned value that is used to establish the probability that an observed trend or difference represents a true difference that is not due to chance alone. If a level of significance is less than the chosen value (usually 1-10%), the difference is regarded as statistically significant. Remember that in rigorous science, it is the hypothesis of no difference or no effect (the null hypothesis, H_0) that is tested. The alternative hypothesis (your tentative explanation for an observation) can only be accepted through statistical rejection of H_0.

Validity: Whether or not you are truly measuring the right thing.

The Scientific Method

Scientific knowledge grows through a process called the **scientific method**. This process involves observation and measurement, hypothesizing and predicting, and planning and executing investigations designed to test formulated **hypotheses**. A scientific hypothesis is a tentative explanation for an observation, which is capable of being tested by experimentation. Hypotheses lead to predictions about the system involved and they are accepted or rejected on the basis of findings arising from the investigation. Rejection of the hypothesis may lead to new, alternative explanations (hypotheses) for the observations. Acceptance of the hypothesis as a valid explanation is not necessarily permanent: explanations may be rejected at a later date in light of new findings. This process eventually leads to new knowledge (theory, laws, or models).

Making Observations

These may involve the observation of certain behaviors in wild populations, physiological measurements made during previous experiments, or 'accidental' results obtained when seeking answers to completely unrelated questions.

Asking Questions

The observations lead to the formation of questions about the system being studied.

Forming a Hypothesis

Features of a sound hypothesis:
- It is based on observations and prior knowledge of the system.
- It offers an explanation for an observation.
- It refers to only one independent variable.
- It is written as a definite statement and not as a question.
- It is testable by experimentation.
- It leads to predictions about the system.

Generating a Null Hypothesis

A hypothesis based on observations is used to generate the **null hypothesis** (H_0); the hypothesis of no difference or no effect. Hypotheses are expressed in the null form for the purposes of statistical testing. H_0 may be rejected in favor of accepting the alternative hypothesis, H_A.

Testing the Predictions

The predictions are tested out in the practical part of an investigation.

Testing predictions may lead to new observations

Accept or reject the hypothesis

Designing an Investigation

Investigations are planned so that the predictions about the system made in the hypothesis can be tested. Investigations may be laboratory or field based.

Making Predictions

Based on a hypothesis, **predictions** (expected, repeatable outcomes) can be generated about the behavior of the system. Predictions may be made on any aspect of the material of interest, e.g. how different variables (factors) relate to each other.

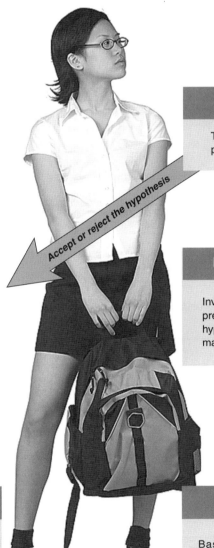

Related activities: Designing Your Experiment

The Differences between Science and Pseudo-Science

The scientific method is rigorous and based on observation, theory, experimentation, documentation, repeatability, and critical review. The underlying principles of science: to observe, question, theorize, and test are quite different from pseudo-science, which begins with a preconceived idea or theory and then looks for certain experimental results or evidence to support that idea, rejecting other contrary evidence in the process. astrology, creation "science", palmistry, and clairvoyancy are examples of this type of pseudo-science. Some of the differences between science and pseudo-science are outlined below:

Science	Pseudo-science
Ability & willingness to change	Fixed ideas
Absorbs all new discoveries	Selected favorable discoveries
Ruthless peer review	No review
Invites critical review	Hostile to criticism
Ability to predict	Inability to predict
Makes testable assertions	Makes non-testable assertions
Verifiable	Non-repeatable ("trust me")
Few assumptions made	Clings to cherished ideas

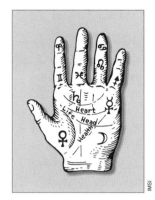

1. (a) A student noticed that woodlice (slaters) were often found in moist areas when she went looking for them. From this observation, she formulated the hypothesis stated below. Describe two features that make this a good hypothesis:

    "Moisture level of the microhabitat influences woodlouse distribution"

    Feature 1: _____

    Feature 2: _____

    (b) Generate a prediction from this hypothesis: _____

2. Generate a prediction for each of the following hypotheses:

    (a) *"The transpiration rate of plant A is influenced by the amount of air movement around it."*

    (b) *"The rate of movement of snail species A up a slope is influenced by the angle of the slope."*

3. Explain why scientific hypotheses are usually expressed in the negative form (i.e. a **null hypothesis**):

4. Discuss the benefits, as you see them, of science over pseudo-science as a way in which to account for the observations we make of the world around us:

A Guide to Research Projects

The following guide will assist you in preparing and executing your own research project. The main points associated with each stage in your project are outlined. Some of these are expanded in detail later in this topic. The right hand column provides a checklist. At the completion of each stage you may wish to consult with your teacher to check your progress.

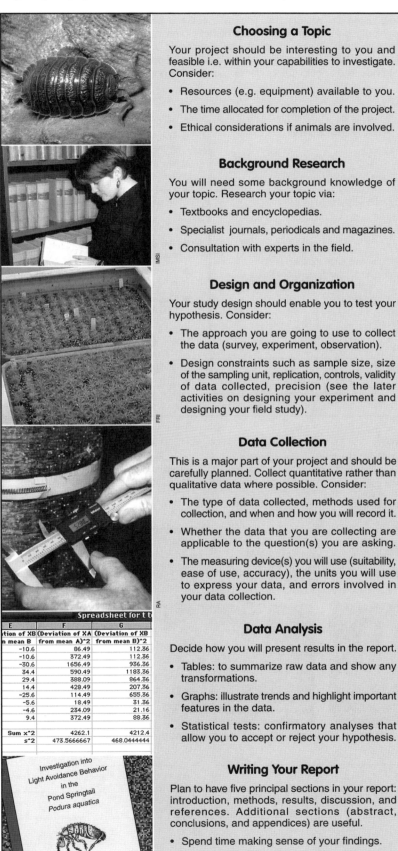

Choosing a Topic

Your project should be interesting to you and feasible i.e. within your capabilities to investigate. Consider:

- Resources (e.g. equipment) available to you.
- The time allocated for completion of the project.
- Ethical considerations if animals are involved.

Background Research

You will need some background knowledge of your topic. Research your topic via:

- Textbooks and encyclopedias.
- Specialist journals, periodicals and magazines.
- Consultation with experts in the field.

Design and Organization

Your study design should enable you to test your hypothesis. Consider:

- The approach you are going to use to collect the data (survey, experiment, observation).
- Design constraints such as sample size, size of the sampling unit, replication, controls, validity of data collected, precision (see the later activities on designing your experiment and designing your field study).

Data Collection

This is a major part of your project and should be carefully planned. Collect quantitative rather than qualitative data where possible. Consider:

- The type of data collected, methods used for collection, and when and how you will record it.
- Whether the data that you are collecting are applicable to the question(s) you are asking.
- The measuring device(s) you will use (suitability, ease of use, accuracy), the units you will use to express your data, and errors involved in your data collection.

Data Analysis

Decide how you will present results in the report.

- Tables: to summarize raw data and show any transformations.
- Graphs: illustrate trends and highlight important features in the data.
- Statistical tests: confirmatory analyses that allow you to accept or reject your hypothesis.

Writing Your Report

Plan to have five principal sections in your report: introduction, methods, results, discussion, and references. Additional sections (abstract, conclusions, and appendices) are useful.

- Spend time making sense of your findings.
- Be clear about what you are going to write about.
- Prepare a draft and revise it. Proof read carefully.

Project Plan Checklist

☐ Project is feasible given available resources.
☐ There is adequate time to complete the study (it is not too ambitious).
☐ Project ethics and animal welfare issues have been addressed.

Teacher's checkpoint: ....

☐ Topic has been researched in enough depth to satisfy the requirements of the project.
☐ An equipment list has been compiled and the availability of the necessary equipment has been assessed.

Teacher's checkpoint: ....

☐ The hypothesis has been clearly stated
☐ The best approach for the study (survey, experimental) has been identified.
☐ If there is a need for a pilot study, this has been identified.
☐ Design features (e.g. sample sizes and controls) meet the needs of the study.
☐ The study is robust enough not to fail completely if one stage goes wrong.

Teacher's checkpoint: ....

Project Execution Checklist

☐ The methods of data collection and the type of data collected are appropriate to the study and the questions being asked.
☐ The methods used for collecting and recording the data (including units of measurement) are consistent throughout.
☐ For safe keeping, there are copies of data and preliminary analyses (if present).

Teacher's checkpoint: ....

☐ Raw data has been summarized and any transformations necessary have been done.
☐ The way in which the data are presented is appropriate for the data involved and the information being conveyed.
☐ Any statistical tests are appropriate to the data and the hypothesis being tested.

Teacher's checkpoint: ....

☐ The report contains the necessary sections, with additional appendices if necessary.
☐ Each section contains material appropriate to that section.
☐ The reference list, in text citations, figures and tables are correctly formatted.
☐ The draft has been proof read.

Teacher's assessment: ....

Related activities: Choosing Your Topic, Designing Your Experiment, Taking the Next Step

Choosing Your Topic

When planning any kind of biological investigation, it is important to consider the type of data that will be collected. It is best, whenever possible, to collect quantitative or numerical data, as these data lend themselves well to analysis and statistical testing. Recording data in a systematic way as you collect it, e.g. using a table or spreadsheet, is important, especially if data manipulation and transformation are required. It is important to calculate summary, descriptive statistics (e.g. mean or median) as you proceed. These will help you to recognise important trends and features in your data as they become apparent. The biggest hurdle in undertaking an experimental study will be in choosing a topic that lends itself to the aims of the investigation and is designed in such a way that analysis is straightforward and biologically meaningful. The case outlined below relates to an investigation of an organism's niche. Some guidelines for consideration are given, together with a synopsis of types of variables. You should be familiar with the qualities of data before you start, as this will help you to develop a well designed investigation.

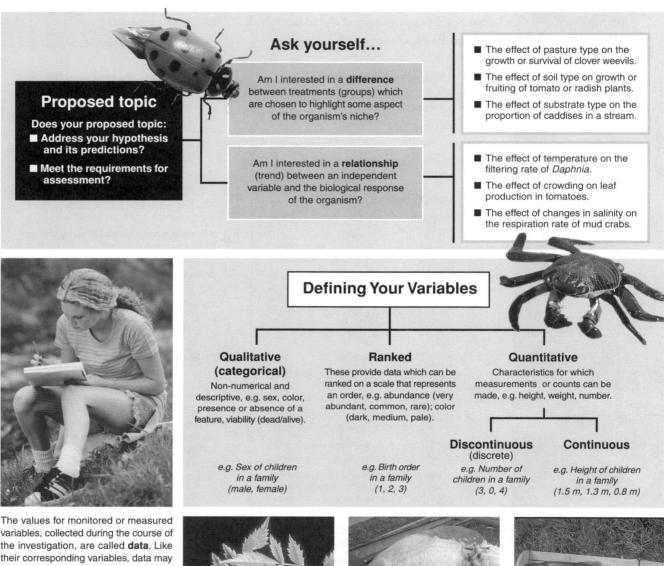

Proposed topic
Does your proposed topic:
- Address your hypothesis and its predictions?
- Meet the requirements for assessment?

Ask yourself...

Am I interested in a **difference** between treatments (groups) which are chosen to highlight some aspect of the organism's niche?

- The effect of pasture type on the growth or survival of clover weevils.
- The effect of soil type on growth or fruiting of tomato or radish plants.
- The effect of substrate type on the proportion of caddises in a stream.

Am I interested in a **relationship** (trend) between an independent variable and the biological response of the organism?

- The effect of temperature on the filtering rate of *Daphnia*.
- The effect of crowding on leaf production in tomatoes.
- The effect of changes in salinity on the respiration rate of mud crabs.

Defining Your Variables

Qualitative (categorical)
Non-numerical and descriptive, e.g. sex, color, presence or absence of a feature, viability (dead/alive).
e.g. Sex of children in a family (male, female)

Ranked
These provide data which can be ranked on a scale that represents an order, e.g. abundance (very abundant, common, rare); color (dark, medium, pale).
e.g. Birth order in a family (1, 2, 3)

Quantitative
Characteristics for which measurements or counts can be made, e.g. height, weight, number.

- **Discontinuous** (discrete)
  e.g. Number of children in a family (3, 0, 4)
- **Continuous**
  e.g. Height of children in a family (1.5 m, 1.3 m, 0.8 m)

The values for monitored or measured variables, collected during the course of the investigation, are called **data**. Like their corresponding variables, data may be quantitative, qualitative, or ranked.

Leaf shape: qualitative

Number per litter: quantitative, discontinuous

Fish length: quantitative, continuous

Use this space to develop some ideas about potential topics. For each, identify the nature of the variables involved:

1. _____

2. _____

3. _____

Hypotheses and Predictions

Generating a **hypothesis** is crucial to scientific investigation. A hypothesis offers a tentative explanation to questions generated by observations. For the purposes of investigation, hypotheses are often constructed in a form that allows them to be tested statistically. For every hypothesis, there is a corresponding **null hypothesis**; a hypothesis against the **prediction**. Predictions are tested with laboratory and field experiments and carefully focused observations. For a hypothesis to be accepted it should be possible for anyone to test the predictions with the same methods and get a similar result each time. Scientific hypotheses may be modified as more information becomes available. In this activity, you will be able to practise making hypotheses and predictions based on a simple predator-prey system about which certain **assumptions** have been made.

Observations, Hypotheses, and Predictions

Observation is the basis for formulating hypotheses and making predictions. An observation may generate a number of plausible hypotheses, and each hypothesis will lead to one or more predictions, which can be tested by further investigation.

Observation 1: Some caterpillar species are brightly colored and appear to be conspicuous to predators such as insectivorous birds. Predators appear to avoid these species. These caterpillars are often found in groups, rather than as solitary animals.

Observation 2: Some caterpillar species are cryptic in their appearance or behavior. Their camouflage is so convincing that, when alerted to danger, they are difficult to see against their background. Such caterpillars are usually found alone.

Assumptions

In any biological investigations, you make certain assumptions about the biological system you are working with. Assumptions are features of the system (and your investigation) that you assume to be true but do not (or cannot) test. Possible assumptions about the biological system described above include:

- Insectivorous birds have color vision.
- Caterpillars that appear brightly colored or cryptic to us, also appear that way to insectivorous birds.
- Insectivorous birds are able to learn about the palatability of prey by tasting them.

Useful Types of Hypotheses

Hypothesis involving manipulation
Used when the effect of manipulating a variable on a biological entity is being investigated.
Example: The composition of applied fertilizer influences the rate of growth of plant A.

Hypothesis of choice
Used when species preference, e.g. for a particular habitat type or microclimate, is being investigated.
Example: Woodpeckers (species A) show a preference for tree type when nesting.

Hypothesis involving observation
Used when organisms are being studied in their natural environment and conditions cannot be changed. **Example**: Fern abundance is influenced by the degree to which the canopy is established.

1. Study the previous example illustrating the features of cryptic and conspicuous caterpillars, then answer the following:

   (a) Generate a hypothesis to explain the observation that some caterpillars are brightly colored and conspicuous while others are cryptic and blend into their surroundings:

   Hypothesis: _____

   _____

   _____

   (b) State the null form of this hypothesis: _____

   _____

   (c) Describe one of the assumptions being made in your hypothesis: _____

   _____

   _____

   (d) Based on your hypothesis, generate a prediction about the behavior of insectivorous birds towards caterpillars:

   _____

   _____

3. During the course of any investigation, new information may arise as a result of observations unrelated to the original hypothesis. This can lead to the generation of further hypotheses about the system. For each of the incidental observations described below, formulate a prediction, and an outline of an investigation to test it. *The observation described in each case was not related to the hypothesis the experiment was designed to test:*

   (a) **Bacterial cultures**

   Prediction: _____

   _____

   _____

   Outline of the investigation: _____

   _____

   _____

   _____

   _____

   _____

   Bacterial Cultures

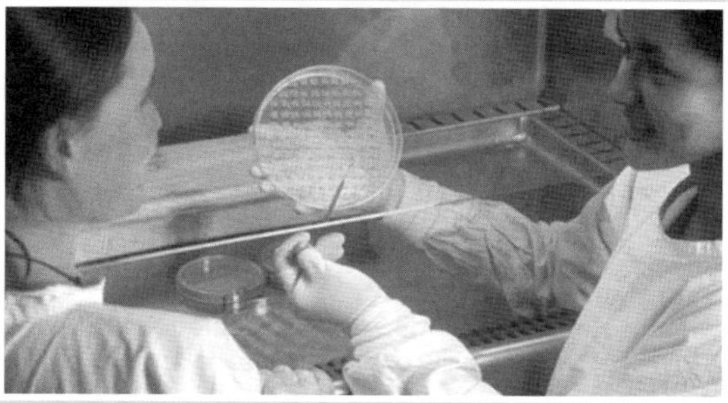

   Observation: During an experiment on bacterial growth, the girls noticed that the cultures grew at different rates when the dishes were left overnight in different parts of the laboratory.

   (b) **Plant cloning**

   Prediction: _____

   _____

   _____

   Outline of the investigation: _____

   _____

   _____

   _____

   _____

   _____

   Plant Cloning

   Observation: During an experiment on plant cloning, a scientist noticed that the root length of plant clones varied depending on the concentration of a hormone added to the agar.

Planning an Investigation

Investigations involve written stages (planning and reporting), at the start and end. The middle stage is the practical work when the data are collected. Practical work may be laboratory or field based. Typical lab based studies involve investigating how a biological response is affected by manipulating a particular **variable**, e.g. temperature. Field work often involves investigating features of a population or community. These may be interrelationships, such as competition, or patterns, such as zonation. Where quantitative information must be gathered from the population or community, particular techniques (such as quadrat sampling) and protocols (e.g. random placement of sampling units) apply. These aspects of practical work are covered in the topic *Field Studies*. Investigations in the field are usually more complex than those in the laboratory because natural systems have many more variables that cannot easily be controlled or accounted for.

Planning

- Formulate your hypothesis from an observation.
- Use a checklist (see the next activity) or a template (above) to construct a plan.

Execution

- Spend time (as appropriate to your study) collecting the data.
- Record the data in a systematic format (e.g. a table or spreadsheet).

Analysis and Reporting

- Analyze the data using graphs, tables, or statistics to look for trends or patterns.
- Write up your report including all the necessary sections.

Identifying Variables

A variable is any characteristic or property able to take any one of a range of values. Investigations often look at the effect of changing one variable on another. It is important to identify all variables in an investigation: independent, dependent, and controlled, although there may be nuisance factors of which you are unaware. In all fair tests, only one variable is changed by the investigator.

Dependent variable
- Measured during the investigation.
- Recorded on the y axis of the graph.

Controlled variables
- Factors that are kept the same or controlled.
- List these in the method, as appropriate to your own investigation.

Independent variable
- Set by the person carrying out the investigation.
- Recorded on the x axis of the graph.

Assumptions

In any experimental work, you will make certain assumptions about the biological system you are working with.

Assumptions are features of the system (and your experiment) that you assume to be true but do not (or cannot) test.

Examples of Investigations

Aim		Variables	
Investigate the effect of varying ...	on the following ...	Independent variable	Dependent variable
Temperature	Leaf width	Temperature	Leaf width
Light intensity	Activity of woodlice	Light intensity	Woodlice activity
Soil pH	Plant height at age 6 months	pH	Plant height

Related activities: A Guide to Research Projects, Choosing Your Topic

In order to write a sound method for your investigation, you need to determine how the independent, dependent, and controlled variables will be set and measured (or monitored). A good understanding of your methodology is crucial to a successful investigation. You need to be clear about how much data, and what type of data, you will collect. You should also have a good idea about how you plan to analyze those data. Use the example below to practise identifying this type of information.

Case Study: Catalase Activity

Catalase is an enzyme that converts hydrogen peroxide (H_2O_2) to oxygen and water. An experiment investigated the effect of temperature on the rate of the catalase reaction. Small (10 cm^3) test tubes were used for the reactions, each containing 0.5 cm^3 of enzyme and 4 cm^3 of hydrogen peroxide. Reaction rates were assessed at four temperatures (10°C, 20°C, 30°C, and 60°C). For each temperature, there were two reaction tubes (e.g. tubes 1 and 2 were both kept at 10°C). The height of oxygen bubbles present after one minute of reaction was used as a measure of the reaction rate; a faster reaction rate produced more bubbles. The entire experiment, involving eight tubes, was repeated on two separate days.

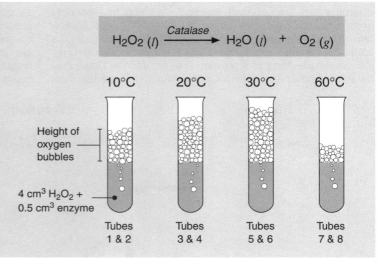

$$H_2O_2 (l) \xrightarrow{Catalase} H_2O (l) + O_2 (g)$$

1. Write a suitable aim for this experiment: _____

2. Write a suitable hypothesis for this experiment: _____

3. (a) Identify the **independent variable**: _____

   (b) State the range of values for the independent variable: _____

   (c) Name the unit for the independent variable: _____

   (d) List the equipment needed to set the independent variable, and describe how it was used: _____

4. (a) Identify the **dependent variable**: _____

   (b) Name the unit for the dependent variable: _____

   (c) List the equipment needed to measure the dependent variable, and describe how it was used: _____

5. (a) Each temperature represents a treatment/sample/trial (circle one):

   (b) State the number of tubes at each temperature: _____

   (c) State the sample size for each treatment: _____

   (d) State how many times the whole investigation was repeated: _____

6. Explain why it would have been desirable to have included an extra tube containing no enzyme: _____

7. Identify three variables that might have been controlled in this experiment, and how they could have been monitored:

   (a) _____

   (b) _____

   (c) _____

8. Explain why controlled variables should be monitored carefully: _____

Designing Your Experiment

The figure below provides an example of a basic experimental design aimed at investigating the effect of pH on the growth of a bog adapted plant species. It is not intended to represent a full methodology, but offers some points to consider. An appropriate analysis for this experiment would be a calculation of the mean for each treatment group, followed by a plot of the data with a calculated measure of spread (e.g. 95% confidence intervals). Depending on trends indicated by the plotted data, either a regression or an ANOVA would be an appropriate analysis (see the activities on measures of spread, regression, and ANOVA for more information on the use of these analyses).

Observation
Two students noticed an abundance of a common plant in a boggy area of land near their school. They tested the soil pH in the area and found it to be quite low (between 4 and 5). Garden soil was about pH 7.

Hypothesis
The plant species is well adapted to grow at low pH and will therefore grow better under acid conditions than under conditions of neutral or alkaline pH. The **null hypothesis** is that there is no difference in the growth of this plant species at low and neutral pH.

Notes on preparation, measurement & analysis
- 50 seeds were germinated on damp blotting paper and, of these, 12 in a similar state of germination (10 mm shoot) were chosen for the experiment.
- Each seedling was weighed to the nearest 0.1 g before planting into each of the 12 test pots.
- An estimate of growth rate per day was calculated from the total wet weight of each plant at the end of 20 days.
- The difference in growth rate between the treatment and the control was tested using a Student t test (n = 6).

Experiment
An experiment was designed to test the prediction that the plants would grow best at low pH. The design is shown here (it is not intended to be a full methodology).

Control of Variables

Fixed (controlled) variables: These are controlled and the same across all treatments
- Lighting regime *(quantity and quality)*
- Age and history of plants
- Type and volume of soil
- Pot size and type *(dimensions, material)*
- Watering regime *(volume day^{-1}, frequency)*

Dependent variable: The biological response
- Plant growth rate (g day^{-1}) calculated from wet weight of entire plants (washed and blotted) after 20 days

Independent variable: This is factor that is being manipulated in the experiment
- pH of the water provided to the plants

Other variables: These are factors that you should be aware of but cannot control
- Genetic variation between plants (uncontrollable but assessed by having six plants per treatment)
- Temperature (all plants received the same room temperature regime but this was not controlled)

Fluorescent strip lighting

| Low pH treatment | Control treatment |

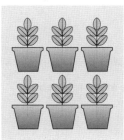 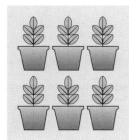

 pH 4 — Watering regime: 100 cm³ per day water at pH 4

 pH 7 — Watering regime: 100 cm³ per day water at pH 7

Assumptions
Features of the experiment that you assume to be true but do not (or cannot) test
- All plants are essentially no different to each other in their growth response at different pH levels
- The soil mix, light quality and quantity and temperature are adequate for healthy continued growth
- Watering volume of 100 cm³ day^{-1} is adequate. This could be tested with a trial experiment beforehand.
- A pH of 7 is a good indicator of the ideal growth pH for most non-acid adapted plants

Plan view of the experimental layout. T = low pH. C = control. Note that in experiments with a large number of treatments and/or replication, it is important to randomize the arrangement of the treatments to account for any effects of location in the set-up.

A note about replication in experiments

Replication refers to the number of times you repeat your entire experimental design (including controls). True replication is not the same as increasing the sample size (n) although it is often used to mean the same thing. Replication accounts for any unusual and unforeseen effects that may be operating in your set-up (e.g. field trials of plant varieties where soil type is variable). Replication is necessary when you expect that the response of treatments will vary because of factors outside your control. It is a feature of higher level experimental designs, and complex statistics are needed to separate differences between replicate treatments. For simple experiments, it is usually more valuable to increase the sample size than to worry about replicates.

1. With reference to the experiment opposite, explain the importance of each of the following:

   (a) Only one plant in each pot: _____

   _____

   (b) The layout of treatments on the bench (see plan view of layout): _____

   _____

2. Explain the best way to take account of the natural variability between individuals when designing an experiment:

   _____

   _____

3. (a) Explain the purpose of replication in experiments: _____

   _____

   (b) Describe factors that would limit your ability to replicate your experimental design: _____

   _____

   (c) Suggest how you could compensate for the lack of true replication: _____

   _____

YOUR CHECKLIST FOR EXPERIMENTAL DESIGN

The following provides a checklist for an experimental design. Check off the points when you are confident that you have satisfied the requirements in each case:

1. **Preliminary:**

   ☐ (a) Makes a hypothesis based on observation(s).

   ☐ (b) The hypothesis (and its predictions) are testable using the resources you have available (the study is feasible).

   ☐ (c) The organism you have chosen is suitable for the study and you have considered the ethics involved.

2. **Variables and assumptions:**

   ☐ (a) You are aware of any assumptions that you are making in your experiment.

   ☐ (b) You have identified all the variables in the experiment (fixed, manipulated, response, uncontrollable).

   ☐ (c) You have considered how you will control the variables you wish to remain fixed and what will be your control.

   ☐ (d) You have considered what preliminary treatment or trials are necessary.

   ☐ (e) You have considered the layout of your treatments to account for any unforeseen variability in your set-up.

3. **Data collection:**

   ☐ (a) You are happy with how and when you are going to take your measurements or samples.

   ☐ (b) You have chosen an appropriate sample size (the number of samples you are going to take).

   ☐ (c) You have given consideration to how you will analyse the data you collect and made sure that your design allows you to answer the questions you wish to answer.

Recording Results

Designing a table to record your results is part of planning your investigation. Once you have collected all your data, you will need to analyze and present it. To do this, it may be necessary to transform your data first, by calculating a mean or a rate. An example of a table for recording results is presented below. This example relates to the investigation described in the previous activity, but it represents a relatively standardised layout. The labels on the columns and rows are chosen to represent the design features of the investigation. The first column contains the entire range chosen for the independent variable. There are spaces for multiple sampling units, repeats (trials), and averages. A version of this table should be presented in your final report.

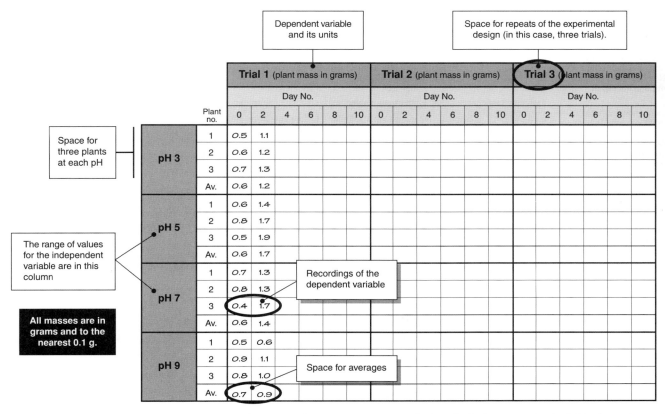

1. In the space (below) design a table to collect data from the case study below. Include space for individual results and averages from the three set ups (use the table above as a guide).

Case Study
Carbon dioxide levels in a respiration chamber

A datalogger was used to monitor the concentrations of carbon dioxide (CO_2) in respiration chambers containing five green leaves from one plant species. The entire study was performed in conditions of full light (quantified) and involved three identical set-ups. The CO_2 concentrations were measured every minute, over a period of ten minutes, using a CO_2 sensor. A mean CO_2 concentration (for the three set-ups) was calculated. The study was carried out two more times, two days apart.

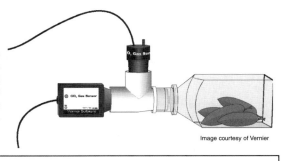

Image courtesy of Vernier

2. Next, the effect of various light intensities (low light, half-light, and full light) on CO_2 concentration was investigated. Describe how the results table for this investigation would differ from the one you have drawn above (for full light only):

_____

_____

Related activities: Transforming Raw Data, Tables and Graphs

DA 2

Tables and Graphs

Data can be presented in a number of ways. Tables provide an accurate record of numerical values and allow you to organise your data in a way that allows you to clarify the relationships and trends that are apparent. Graphical presentation provides a visual image of trends in the data in a minimum of space. The choice between graphing or tabulation depends on the type and complexity of the data and the information that you are wanting to convey. Outlined below are some of the basic rules for constructing tables and graphs. Values for standard deviation are included in this example, although it is not a requirement to calculate these. In your report, always allow enough space for a graph, e.g. one third to one half of an A4 page. The examples in this workbook are usually reduced for reasons of space.

Presenting Data in Tables

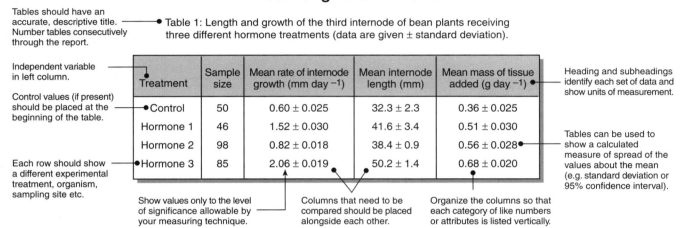

Tables should have an accurate, descriptive title. Number tables consecutively through the report.

Table 1: Length and growth of the third internode of bean plants receiving three different hormone treatments (data are given ± standard deviation).

Independent variable in left column.

Control values (if present) should be placed at the beginning of the table.

Each row should show a different experimental treatment, organism, sampling site etc.

Heading and subheadings identify each set of data and show units of measurement.

Tables can be used to show a calculated measure of spread of the values about the mean (e.g. standard deviation or 95% confidence interval).

Show values only to the level of significance allowable by your measuring technique.

Columns that need to be compared should be placed alongside each other.

Organize the columns so that each category of like numbers or attributes is listed vertically.

Presenting Data in Graph Format

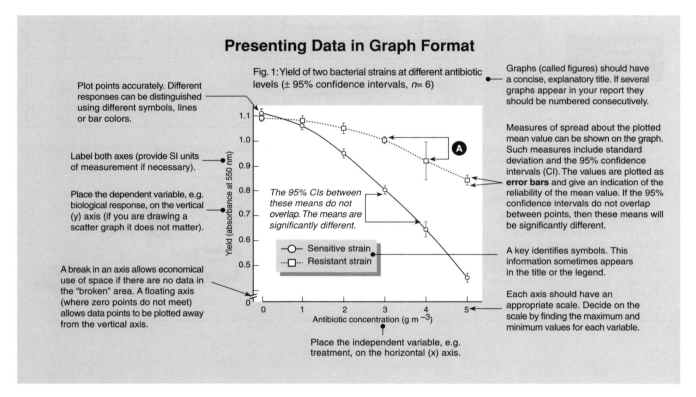

Fig. 1: Yield of two bacterial strains at different antibiotic levels (± 95% confidence intervals, $n = 6$)

Plot points accurately. Different responses can be distinguished using different symbols, lines or bar colors.

Label both axes (provide SI units of measurement if necessary).

Place the dependent variable, e.g. biological response, on the vertical (y) axis (if you are drawing a scatter graph it does not matter).

A break in an axis allows economical use of space if there are no data in the "broken" area. A floating axis (where zero points do not meet) allows data points to be plotted away from the vertical axis.

The 95% CIs between these means do not overlap. The means are significantly different.

Graphs (called figures) should have a concise, explanatory title. If several graphs appear in your report they should be numbered consecutively.

Measures of spread about the plotted mean value can be shown on the graph. Such measures include standard deviation and the 95% confidence intervals (CI). The values are plotted as **error bars** and give an indication of the reliability of the mean value. If the 95% confidence intervals do not overlap between points, then these means will be significantly different.

A key identifies symbols. This information sometimes appears in the title or the legend.

Each axis should have an appropriate scale. Decide on the scale by finding the maximum and minimum values for each variable.

Place the independent variable, e.g. treatment, on the horizontal (x) axis.

1. What can you conclude about the difference (labeled A) between the two means plotted above? Explain your answer:

   _____

2. Discuss the reasons for including both graphs and tables in a final report: _____

   _____

   _____

   _____

Transforming Raw Data

The data collected by measuring or counting in the field or laboratory are called **raw data**. They often need to be changed (**transformed**) into a form that makes it easier to identify important features of the data (e.g. trends). The calculation of **rate** is an example of a commonly performed data transformation, and is appropriate when studying the growth of an organism (or population). If the variable measured was plant height, it could be transformed into a growth rate by considering the time taken for the increase in height to occur. For a line graph with time as the independent variable plotted against the values of the biological response, the slope of the line is a measure of the rate. Biological investigations often compare the rates of events in different situations (e.g. the rate of photosynthesis in the light compared to the rate of photosynthesis in the dark). Typical transformations used in field based investigations involve percentages, frequencies, and totals.

Transformation	Rationale for Transformation
Frequency table	A tally chart of the number of times a value occurs in a data set. It is a useful first step in data analysis as a neatly constructed tally chart can double as a simple histogram.
Total	The sum of all data values for a variable. Useful as an initial stage in data handling, especially in comparing replicates. Used in making other data transformations.
Percentages	Provide a clear expression of what proportion of data fall into any particular category. This relationship may not be obvious from the raw data values.
Rates	Expressed as a measure per unit time. Rates show how a variable changes over a standard time period (e.g. one second, one minute or one hour). Rates allow meaningful comparison of data that may have been recorded over different time periods.
Reciprocals	Reciprocals of time (1/data value) can provide a crude measure of rate in situations where the variable measured is the total time taken to complete a task e.g. time taken for a color change to occur in an enzyme reaction.
Relative values	These involve expression of data values relative to a standard value e.g. number of road deaths per 1000 cars or calorie consumption per gram of body weight. They allow data from different sample sizes or different organisms to be meaningfully compared. Sometimes they are expressed as a percentage (e.g. 35%) or as a proportion (e.g. 0.35).

1. (a) Explain what it means to transform data: _____

    _____

    (b) Briefly explain the general purpose of transforming data: _____

    _____

2. For each of the following examples, state a suitable transformation, together with a reason for your choice:

    (a) Determining relative abundance from counts of four plant species in two different habitat areas:

    Suitable transformation: _____

    Reason: _____

    (b) Making a meaningful comparison between animals of different size in the volume of oxygen each consumed:

    Suitable transformation: _____

    Reason: _____

    (c) Making a meaningful comparison of the time taken for chemical precipitation to occur in a flask at different pH values:

    Suitable transformation: _____

    Reason: _____

    (d) Determining the effect of temperature on the production of carbon dioxide by respiring seeds:

    Suitable transformation: _____

    Reason: _____

3. Complete the transformations for each of the tables on the right. The first value is provided in each case.

(a) TABLE: *Incidence of cyanogenic clover in different areas*

Working: 124 ÷ 159 = 0.78 = 78%

This is the number of cyanogenic clover out of the total.

Incidence of cyanogenic clover in different areas

Clover plant type	Frost free area		Frost prone area		Totals
	Number	%	Number	%	
Cyanogenic	124	78	26		
Acyanogenic	35		115		
Total	159				

(b) TABLE: *Plant transpiration loss using a bubble potometer*

Working: (9.0 − 8.0) ÷ 5 min = 0.2

This is the distance the bubble moved over the first 5 minutes. Note that there is no data entry possible for the first reading (0 min) because no difference can be calculated.

Plant transpiration loss using a bubble potometer

Time (min)	Pipette arm reading (cm^3)	Plant water loss (cm^3 min^{-1})
0	9.0	–
5	8.0	0.2
10	7.2	
15	6.2	
20	4.9	

(c) TABLE: *Photosynthetic rate at different light intensities*

Working: 1 ÷ 15 = 0.067

This is time taken for the leaf to float. A reciprocal gives a per minute rate (the variable measured is the time taken for an event to occur).

NOTE: In this experiment, the flotation time is used as a crude measure of photosynthetic rate. As oxygen bubbles are produced as a product of photosynthesis, they stick to the leaf disc and increase its buoyancy. The faster the rate, the sooner they come to the surface. The rates of photosynthesis should be measured over similar time intervals, so the rate is transformed to a 'per minute' basis (the reciprocal of time).

Photosynthetic rate at different light intensities

Light intensity %	Average time for leaf disc to float (min)	Reciprocal of time (min^{-1})
100	15	0.067
50	25	
25	50	
11	93	
6	187	

(d) TABLE: *Frequency of size classes in a sample of eels*

Working: (7 ÷ 270) × 100 = 2.6 %

This is the number of individuals out of the total that appear in the size class 0-50 mm. The relative frequency is rounded to one decimal place.

Frequency of size classes in a sample of eels

Size class (mm)	Frequency	Relative frequency (%)
0-50	7	2.6
50-99	23	
100-149	59	
150-199	98	
200-249	50	
250-299	30	
300-349	3	
Total	270	

Types of Graphs

Graphs can be used to display data in a way that makes it easy to see trends or relationships between different variables. Before representing data graphically, it is important to identify the kind of data you have. Choosing the correct type of graph can help to clarify an otherwise complicated set of data. It may also highlight important aspects of a relationship between two measured variables. Choosing the wrong type of graph can obscure information and make data more difficult to interpret. Examples of the common types of graph, and the data that are appropriate to them, are provided here. Graphs appropriate for continuous data (line and scatter graphs) are illustrated on this page. Graphs for discontinuous data are illustrated on the following page.

Scatter Graph

In scatter graphs, there is no manipulated (independent) variable but the variables are usually *correlated* i.e. they vary together in some predictable way. The points on the graph need not be connected, but a line of best fit is often drawn through the points (this may be drawn be eye or computer generated).

- The data for this graph must be continuous for both variables
- Useful for determining the relationship between two variables

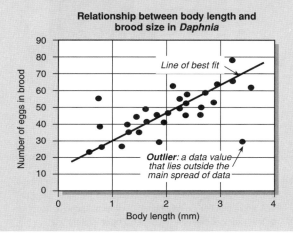

Line Graph

Line graphs are used when one variable (called the independent variable), affects another, the dependent variable. The independent variable is often time or the experimental treatment (controlled variable). The dependent variable is usually the biological response.

- The data for line graphs must be continuous for both variables
- If extreme points are likely to be important, draw a line connecting the points. If there is an implied relationship in the data (e.g. growth increases), a line of best fit that excludes outliers is best.

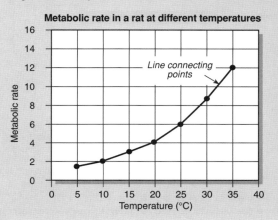

Line Graph with Measure of Spread

Line graphs may be drawn with **error bars**. These are calculated values and indicate the degree of spread (scatter) of the data about a mean. Error bars are used if there are calculated mean (average) values **and** a measure of data spread (e.g. standard deviation).

- Where error bars are large, the data are less consistent (more variable) than in cases where the error bars are small.
- Where no error value has been calculated, the scatter can be shown by plotting the individual data points vertically above and below the mean. By convention, bars are not used to indicate the range of raw values in a data set.

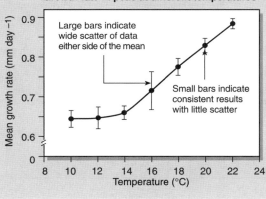

Line Graph: Two Curves Plotted Together

More than one curve can be plotted per set of axes. This is useful when you wish to compare two data sets together.

- If the two data sets use the same measurement units and a similar range of values use one scale and distinguish the two curves with a key
- If the two data sets use different units and/or have a very different range of values use two scales (see example below)
- Adjust scales if necessary to avoid overlapping plots

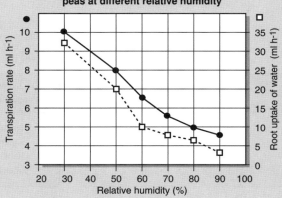

1. (a) Explain what is meant by an outlier: _____

   (b) Suggest how outliers might arise in a data set: _____

   (c) Suggest why outliers are sometimes excluded from an analysis of results: _____

Bar Graph

The data for this graph are non-numerical and **discrete** for at least one variable i.e. they are grouped into separate categories. There are no dependent or independent variables. Axes may be reversed to give graph with the categories on the x axis.

- The data are discontinuous, so the bars do not touch
- Data values may be entered on or above the bars
- Multiple data sets can be displayed using different colored bars placed side by side within the same category

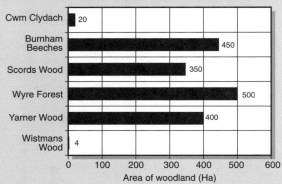

Histograms

Histograms are plots of **continuous** data, usually of some physical variable against frequency of occurrence. Column graphs are drawn to plot frequency distributions when the data are discrete, numerical values (1, 2, 3 etc.). In this case the bars do not touch.

- Histogram data are continuous so the bars touch
- The X-axis usually records the class interval. The Y-axis usually records the number of individuals in each class interval (frequency)

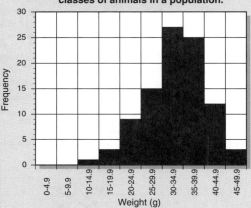

Pie Graph

As with bar graphs, pie graphs are used when the data for one variable are discrete (categories) and the data for the other variable are in the form of counts. A circle is divided according to the proportion of counts in each category. Pie graphs are:

- Good for visual impact and showing relative proportions
- Useful for six or fewer categories
- Not suitable for data sets with a very large number of categories

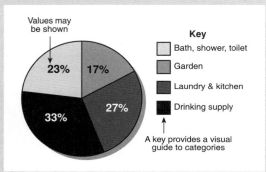

Kite Graph

Kite graphs are elongated figures drawn along a baseline. They can show species distribution across a landscape (or, in some cases, the distribution of one species at different times).

- Each kite represents species abundance across a landscape
- The actual abundance can be calculated from the kite width
- A single thin line on a kite graph represents species absence
- The axes can be reversed depending on preference

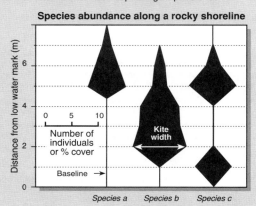

2. Decide which type of graph you would use to plot the data below and give a brief reason for your choice in each case:

Temperature (°C)	Oxygen content of water (mg l⁻¹)
5	9.0
15	7.0
25	5.8
35	5.0

Organism (at sea level)	RBC (millions ml⁻¹ blood)
Human	5.0
Sheep	10.5
Rabbit	4.55
Vicuna	14.9

Size of leaf (mm)	Number of individuals
15 - 19	5
20 - 24	11
25 - 29	16
30 - 34	6

(a) Graph type: _____ Reason: _____

(b) Graph type: _____ Reason: _____

(c) Graph type: _____ Reason: _____

3. Explain why it is important to choose an appropriate class interval when constructing a histogram:

Drawing Bar Graphs

Guidelines for Bar Graphs

Bar graphs are appropriate for data that are non-numerical and **discrete** for at least one variable, i.e. they are grouped into separate categories. There are no dependent or independent variables. Important features of this type of graph include:

- Data are collected for discontinuous, non-numerical categories (e.g. place, color, and species), so the bars do not touch.
- Data values may be entered on or above the bars if you wish.
- Multiple sets of data can be displayed side by side for direct comparison (e.g. males and females in the same age group).
- Axes may be reversed so that the categories are on the x axis, i.e. the bars can be vertical or horizontal. When they are vertical, these graphs are sometimes called column graphs.

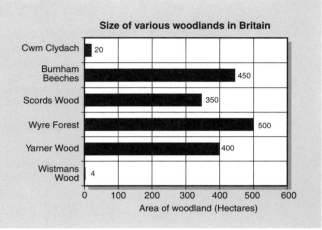

1. Counts of eight mollusc species were made from a series of quadrat samples at two sites on a rocky shore. The summary data are presented here.

   (a) Tabulate the mean (**average**) numbers per square meter at each site in the table below.

   (b) Plot a **bar graph** of the tabulated data on the grid below. For each species, plot the data from both sites side by side using different colors to distinguish the sites.

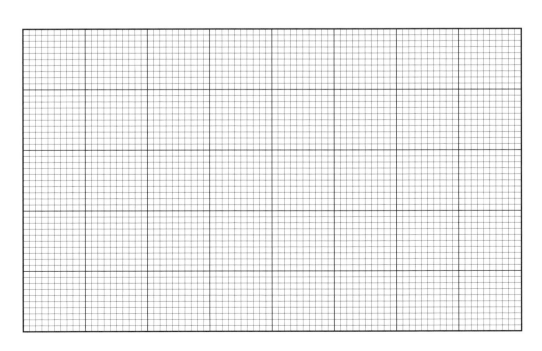

Average abundance of 8 molluscan species from two sites along a rocky shore.

Species	Mean (no. m^{-2})	
	Site 1	Site 2

Drawing Histograms

Guidelines for Histograms

Histograms are plots of **continuous** data and are often used to represent frequency distributions, where the y-axis shows the number of times a particular measurement or value was obtained. For this reason, they are often called frequency histograms. Important features of this type of graph include:

- The data are numerical and continuous (e.g. height or weight), so the bars touch.

- The x-axis usually records the class interval. The y-axis usually records the number of individuals in each class interval (frequency).

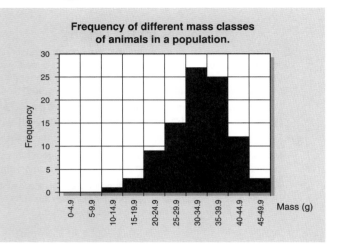

1. The weight data provided below were recorded from 95 individuals (male and female), older than 17 years.

    (a) Create a tally chart (frequency table) in the frame provided, organizing the weight data into a form suitable for plotting. An example of the tally for the weight grouping 55-59.9 kg has been completed for you as an example. Note that the raw data values, once they are recorded as counts on the tally chart, are crossed off the data set in the notebook. It is important to do this in order to prevent data entry errors.

    (b) Plot a **frequency histogram** of the tallied data on the grid provided below.

Drawing Pie Graphs

Guidelines for Pie Graphs

Pie graphs can be used instead of bar graphs, generally in cases where there are six or fewer categories involved. A pie graph provides strong visual impact of the relative proportions in each category, particularly where one of the categories is very dominant. Features of pie graphs include:

- The data for one variable are discontinuous (non-numerical or categories).
- The data for the dependent variable are usually in the form of counts, proportions, or percentages.
- Pie graphs are good for visual impact and showing relative proportions.
- They are not suitable for data sets with a large number of categories.

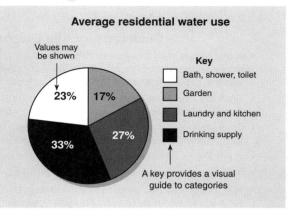

Average residential water use

1. The data provided below are from a study of the diets of three vertebrates.

   (a) Tabulate the data from the notebook shown. Calculate the angle for each percentage, given that each percentage point is equal to 3.6° (the first example is provided: 23.6 x 3.6 = 85).

   (b) Plot a pie graph for each animal in the circles provided. The circles have been marked at 5° intervals to enable you to do this exercise without a protractor. For the purposes of this exercise, begin your pie graphs at the 0° (= 360°) mark and work in a clockwise direction from the largest to the smallest percentage. Use one key for all three pie graphs.

Field data notebook
% of different food items in the diet

Food item	Ferrets	Rats	Cats
Birds	23.6	1.4	6.9
Crickets	15.3	23.6	0
Other insects (not crickets)	15.3	20.8	1.9
Voles	9.2	0	19.4
Rabbits	8.3	0	18.1
Rats	6.1	0	43.1
Mice	13.9	0	10.6
Fruits and seeds	0	40.3	0
Green leaves	0	13.9	0
Unidentified	8.3	0	0

Percentage occurrence of different foods in the diet of ferrets, rats, and cats. Graph angle representing the % is shown to assist plotting.

Food item in diet	Ferrets		Rats		Cats	
	% in diet	Angle (°)	% in diet	Angle (°)	% in diet	Angle (°)
Birds	23.6	85				

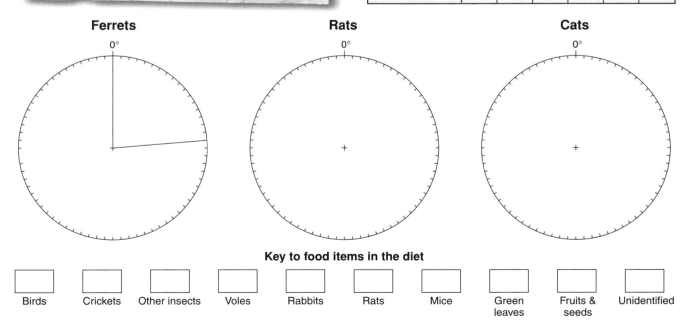

Key to food items in the diet

Birds | Crickets | Other insects | Voles | Rabbits | Rats | Mice | Green leaves | Fruits & seeds | Unidentified

Drawing Kite Graphs

Guidelines for Kite Graphs

Kite graphs are ideal for representing distributional data, e.g. abundance along an environmental gradient. They are elongated figures drawn along a baseline. Important features of kite graphs include:

- Each kite represents changes in species abundance across a landscape. The abundance can be calculated from the kite width.
- They often involve plots for more than one species; this makes them good for highlighting probable differences in habitat preferences between species.
- A thin line on a kite graph represents species absence.
- The axes can be reversed depending on preference.
- Kite graphs may also be used to show changes in distribution with time, for example, with daily or seasonal cycles of movement.

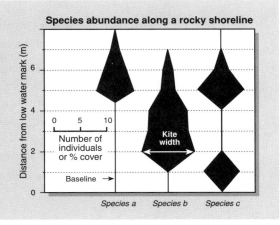

1. The following data were collected from three streams of different lengths and flow rates. Invertebrates were collected at 0.5 km intervals from the headwaters (0 km) to the stream mouth. Their wet weight was measured and recorded (per m^2).

   (a) Tabulate the data below for plotting.

   (b) Plot a **kite graph** of the data from all three streams on the grid provided below. Do not forget to include a scale so that the weight at each point on the kite can be calculated.

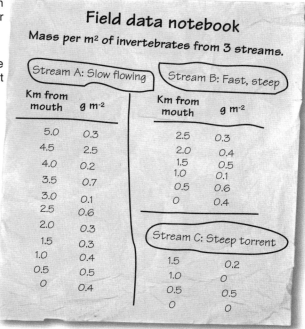

Wet mass of invertebrates along three different streams

Distance from mouth (km)	Wet weight (g m⁻²)		
	Stream A	Stream B	Stream C

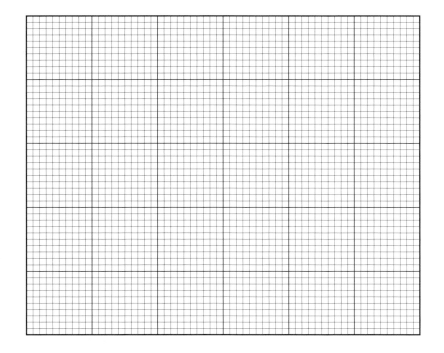

DA 2 — Related activities: Types of Graphs

© Biozone International 2006-2007
Photocopying Prohibited

Drawing Scatter Plots

Guidelines for Scatter Graphs

A scatter graph is a common way to display continuous data where there is a relationship between two interdependent variables.

- The data for this graph must be continuous for both variables.
- There is no independent (manipulated) variable, but the variables are often correlated, i.e. they vary together in some predictable way.
- Scatter graphs are useful for determining the relationship between two variables.
- The points on the graph need not be connected, but a line of best fit is often drawn through the points to show the relationship between the variables (this may be drawn be eye or computer generated).

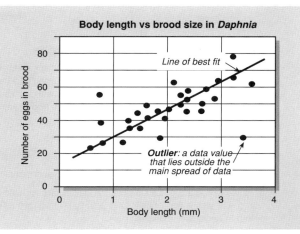

Body length vs brood size in *Daphnia*

1. In the example below, metabolic measurements were taken from seven Antarctic fish *Pagothenia borchgrevinski*. The fish are affected by a gill disease, which increases the thickness of the gas exchange surfaces and affects oxygen uptake. The results of oxygen consumption of fish with varying amounts of affected gill (at rest and swimming) are tabulated below.

    (a) Using **one** scale only for oxygen consumption, plot the data on the grid below to show the relationship between oxygen consumption and the amount of gill affected by disease. Use different symbols or colors for each set of data (at rest and swimming).

    (b) Draw a line of best fit through each set of points.

2. Describe the relationship between the amount of gill affected and oxygen consumption in the fish:

    (a) For the **at rest** data set:

    (b) For the **swimming** data set:

Oxygen consumption of fish with affected gills

Fish number	Percentage of gill affected	Oxygen consumption ($cm^3\ g^{-1}\ h^{-1}$)	
		At rest	Swimming
1	0	0.05	0.29
2	95	0.04	0.11
3	60	0.04	0.14
4	30	0.05	0.22
5	90	0.05	0.08
6	65	0.04	0.18
7	45	0.04	0.20

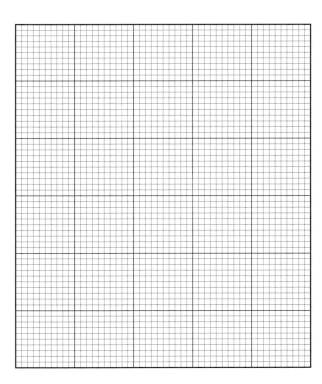

3. Describe how the gill disease affects oxygen uptake in resting fish:

Related activities: Types of Graphs, Interpreting Line Graphs, Linear Regression

DA 2

Drawing Line Graphs

Guidelines for Line Graphs

Line graphs are used when one variable (the independent variable) affects another, the dependent variable. Line graphs can be drawn without a measure of spread (top figure, right) or with some calculated measure of data variability (bottom figure, right). Important features of line graphs include:

- The data must be continuous for both variables.
- The dependent variable is usually the biological response.
- The independent variable is often time or the experimental treatment.
- In cases where there is an implied trend (e.g. one variable increases with the other), a line of best fit is usually plotted through the data points to show the relationship.
- If fluctuations in the data are likely to be important (e.g. with climate and other environmental data) the data points are usually connected directly (point to point).
- Line graphs may be drawn with measure of error. The data are presented as points (the calculated means), with bars above and below, indicating a measure of variability or spread in the data (e.g. standard error, standard deviation, or 95% confidence intervals).
- Where no error value has been calculated, the scatter can be shown by plotting the individual data points vertically above and below the mean. By convention, bars are not used to indicate the range of raw values in a data set.

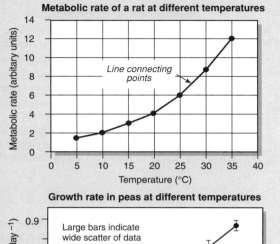

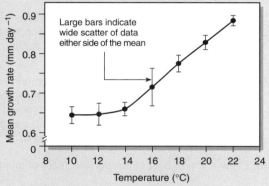

1. The results (shown right) were collected in a study investigating the effect of temperature on the activity of an enzyme.

   (a) Using the results provided in the table (right), plot a line graph on the grid below:

   (b) Estimate the rate of reaction at 15°C: _____

Lab Notebook

An enzyme's activity at different temperatures

Temperature (°C)	Rate of reaction (mg of product formed per minute)
10	1.0
20	2.1
30	3.2
35	3.7
40	4.1
45	3.7
50	2.7
60	0

Plotting Multiple Data Sets

A single figure can be used to show two or more data sets, i.e. more than one curve can be plotted per set of axes. This type of presentation is useful when you want to visually compare the trends for two or more treatments, or the response of one species against the response of another. Important points regarding this format are:

- If the two data sets use the same measurement units and a similar range of values for the independent variable, one scale on the y axis is used.
- If the two data sets use different units and/or have a very different range of values for the independent variable, two scales for the y axis are used (see example provided). The scales can be adjusted if necessary to avoid overlapping plots.
- The two curves must be distinguished with a key.

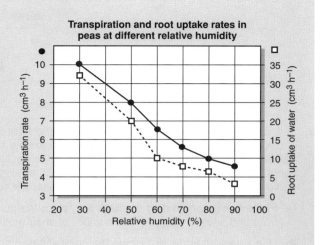

Transpiration and root uptake rates in peas at different relative humidity

2. A census of a deer population on an island indicated a population of 2000 animals in 1960. In 1961, ten wolves (natural predators of deer) were brought to the island in an attempt to control deer numbers. Over the next nine years, the numbers of deer and wolves were monitored. The results of these population surveys are presented in the table, right.

(a) Plot a line graph (joining the data points) for the tabulated results. Use one scale (on the left) for numbers of deer and another scale (on the right) for the number of wolves. Use different symbols or colors to distinguish the lines and include a key.

Field data notebook
Results of a population survey on an island

Time (yr)	Wolf numbers	Deer numbers
1961	10	2000
1962	12	2300
1963	16	2500
1964	22	2360
1965	28	2244
1966	24	2094
1967	21	1968
1968	18	1916
1969	19	1952

(b) Study the line graph that you plotted for the wolf and deer census on the previous page. Provide a plausible explanation for the pattern in the data, stating the evidence available to support your reasoning:

_____

_____

_____

_____

3. In a sampling program, the number of perch and trout in a hydro-electric reservoir were monitored over a period of time. A colony of black shag was also present. Shags take large numbers of perch and (to a lesser extent) trout. In 1960-61, 424 shags were removed from the lake during the nesting season and nest counts were made every spring in subsequent years. In 1971, 60 shags were removed from the lake, and all existing nests dismantled. The results of the population survey are tabulated below (for reasons of space, the entire table format has been repeated to the right for 1970-1978).

(a) Plot a line graph (joining the data points) for the survey results. Use one scale (on the left) for numbers of perch and trout and another scale for the number of shag nests. Use different symbols to distinguish the lines and include a key.

(b) Use a vertical arrow to indicate the point at which shags and their nests were removed.

Results of population survey at a reservoir

Time (yr)	Fish number (average per haul)		Shag nest numbers	Time (yr) continued	Fish number (average per haul)		Shag nest numbers
	Trout	Perch			Trout	Perch	
1960	–	–	16	1970	1.5	6	35
1961	–	–	4	1971	0.5	0.7	42
1962	1.5	11	5	1972	1	0.8	0
1963	0.8	9	10	1973	0.2	4	0
1964	0	5	22	1974	0.5	6.5	0
1965	1	1	25	1975	0.6	7.6	2
1966	1	2.9	35	1976	1	1.2	10
1967	2	5	40	1977	1.2	1.5	32
1968	1.5	4.6	26	1978	0.7	2	28
1969	1.5	6	32				

Source: Data adapted from 1987 Bursary Examination

Interpreting Line Graphs

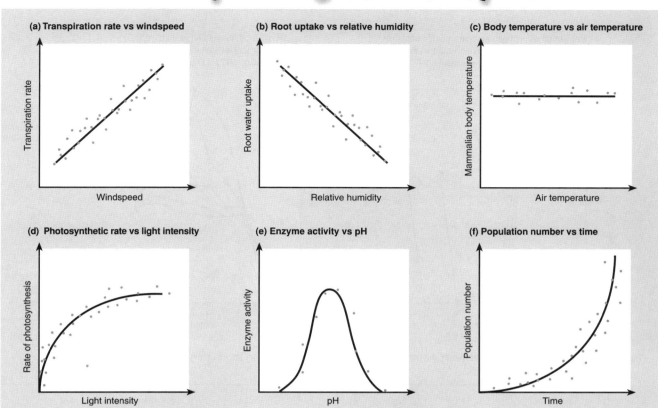

1. For each of the graphs (b-f) above, give a description of the slope and an interpretation of how one variable changes with respect to the other. For the purposes of your description, call the independent variable (horizontal or x-axis) in each example "variable X" and the dependent variable (vertical or y-axis) "variable Y". Be aware that the existence of a relationship between two variables does not necessarily mean that the relationship is causative (although it may be).

   (a) Slope: _Positive linear relationship, with constantly rising slope_

   Interpretation: _Variable Y (transpiration) increases regularly with increase in variable X (windspeed)_

   (b) Slope: _____

   Interpretation: _____

   (c) Slope: _____

   Interpretation: _____

   (d) Slope: _____

   Interpretation: _____

   (e) Slope: _____

   Interpretation: _____

   (f) Slope: _____

   Interpretation: _____

2. Study the line graph of trout, perch and shag numbers that you plotted on the previous page:

   (a) Describe the evidence suggesting that the shag population is exercising some control over perch numbers:

   _____

   (b) Describe evidence that the fluctuations in shag numbers are related to fluctuations in trout numbers:

   _____

3. A survey of two species of bottom dwelling bloodworm was carried out in a lake (for the purposes of this exercise, this is a hypothetical lake). The two species were *Chironomus species a*, and *Chironomus species b*. The water temperature and dissolved oxygen found at various depths in the lake throughout the year are given in the graphs below:

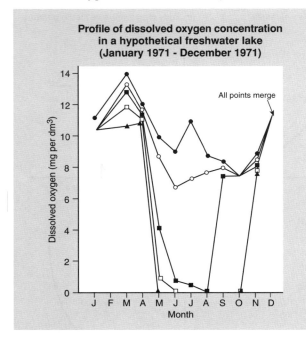
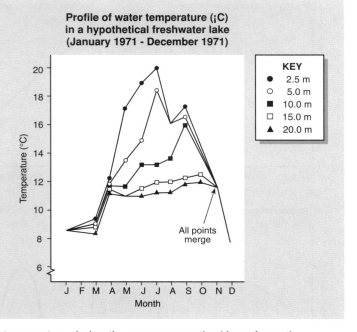

(a) Describe the relationship between water depth and temperature during the summer months (June-August):

_____

_____

(b) Describe the relationship between water depth and dissolved oxygen during the summer months:

_____

_____

4. The column graphs on the right show depth distributions of similar aged individuals of the two bloodworm species described above.

(a) Describe the difference between summer (June-August) and winter (Dec-Feb) distributions. Suggest a reason for this:

_____

_____

_____

_____

_____

(b) Explain why the population density of both species is higher at greater depth during December:

_____

_____

_____

_____

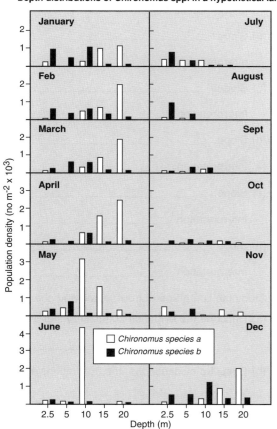

© Biozone International 2006-2007
Photocopying Prohibited

Taking the Next Step

By this stage, you will have completed many of the early stages of your investigation. Now is a good time to review what you have done and reflect on the biological significance of what you are investigating. Review the first page of this flow chart in light of your findings so far. You are now ready to begin a more in-depth analysis of your results. Never under-estimate the value of plotting your data, even at a very early stage. This will help you decide on the best type of data analysis (see next page).

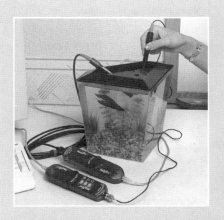

Photos courtesy of Pasco

Observation

Something ...

- Changes or affects something else.
- Is more abundant, etc. along a transect, at one site, temperature, concentration, etc. than others.
- Is bigger, taller, or grows more quickly.

Pilot study

Lets you check ...

- Equipment, sampling sites, sampling interval.
- How long it takes to collect data.
- Problems with identification or other unforeseen issues.

Research

To find out ...

- Basic biology and properties.
- What other biotic or abiotic factors may have an effect.
- Its place within the broader biological context.

Analysis

Are you looking for a ...

- **Difference**.
- **Trend** or relationship.
- **Goodness of fit** (to a theoretical outcome).

GO TO NEXT PAGE

Be prepared to revise your study design in the light of the results from your pilot study

Variables

Next you need to ...

- Identify the key variables likely to cause the effect.
- Identify variables to be controlled in order to give the best chance of showing the effect that you want to study.

Hypothesis

Must be ...

- Testable
- Able to generate predictions

so that in the end you can say whether your data supports or allows you to reject your hypothesis.

© Biozone International 2006-2007
Photocopying Prohibited

Related activities: Linear Regression, Non-linear Regression, Student's *t*-Test, Chi-Squared Test, Chi-Squared Exercise in Ecology, Analysis of Variance

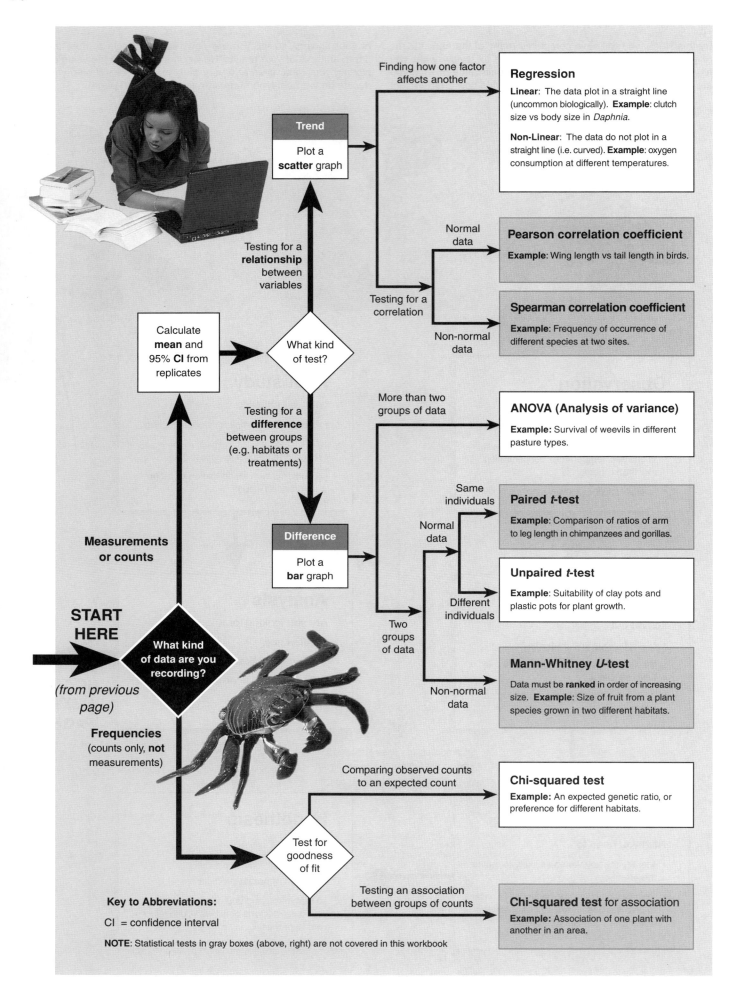

Descriptive Statistics

For most investigations, measures of the biological response are made from more than one sampling unit. The sample size (the number of sampling units) will vary depending on the resources available. In lab based investigations, the sample size may be as small as two or three (e.g. two test-tubes in each treatment). In field studies, each individual may be a sampling unit, and the sample size can be very large (e.g. 100 individuals). It is useful to summarize the data collected using **descriptive statistics**.

Descriptive statistics, such as mean, median, and mode, can help to highlight trends or patterns in the data. Each of these statistics is appropriate to certain types of data or distributions, e.g. a mean is not appropriate for data with a skewed distribution (see below). Frequency graphs are useful for indicating the distribution of data. Standard deviation and standard error are statistics used to quantify the amount of spread in the data and evaluate the reliability of estimates of the true (population) mean.

Variation in Data

Whether they are obtained from observation or experiments, most biological data show variability. In a set of data values, it is useful to know the value about which most of the data are grouped; the center value. This value can be the mean, median, or mode depending on the type of variable involved (see schematic below). The main purpose of these statistics is to summarize important trends in your data and to provide the basis for statistical analyses.

Distribution of Data

Variability in continuous data is often displayed as a **frequency distribution**. A frequency plot will indicate whether the data have a normal distribution (A), with a symmetrical spread of data about the mean, or whether the distribution is skewed (B), or bimodal (C). The shape of the distribution will determine which statistic (mean, median, or mode) best describes the central tendency of the sample data.

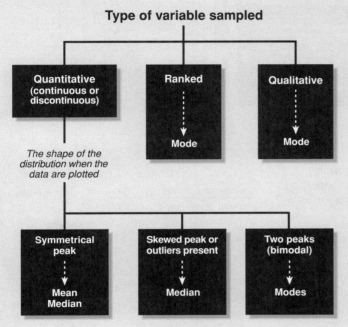

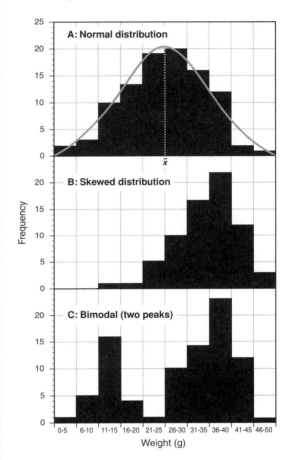

Statistic	Definition and use	Method of calculation
Mean	• The average of all data entries. • Measure of central tendency for normally distributed data.	• Add up all the data entries. • Divide by the total number of data entries.
Median	• The middle value when data entries are placed in rank order. • A good measure of central tendency for skewed distributions.	• Arrange the data in increasing rank order. • Identify the middle value. • For an even number of entries, find the mid point of the two middle values.
Mode	• The most common data value. • Suitable for bimodal distributions and qualitative data.	• Identify the category with the highest number of data entries using a tally chart or a bar graph.
Range	• The difference between the smallest and largest data values. • Provides a crude indication of data spread.	• Identify the smallest and largest values and find the difference between them.

When NOT to calculate a mean:

In certain situations, calculation of a simple arithmetic mean is inappropriate.

Remember:

- *DO NOT* calculate a mean from values that are already means (averages) themselves.
- *DO NOT* calculate a mean of ratios (e.g. percentages) for several groups of different sizes; go back to the raw values and recalculate.
- *DO NOT* calculate a mean when the measurement scale is not linear, e.g. pH units are not measured on a linear scale.

Measuring Spread

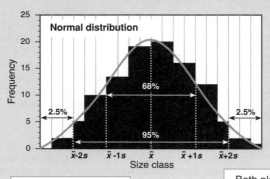

The **standard deviation** is a frequently used measure of the variability (spread) in a set of data. It is usually presented in the form $\bar{x} \pm s$. In a normally distributed set of data, 68% of all data values will lie within one standard deviation (s) of the mean ($\bar{x}$) and 95% of all data values will lie within two standard deviations of the mean (left).

Two different sets of data can have the same mean and range, yet the distribution of data within the range can be quite different. In both the data sets pictured in the histograms below, 68% of the values lie within the range $\bar{x} \pm 1s$ and 95% of the values lie within $\bar{x} \pm 2s$. However, in B, the data values are more tightly clustered around the mean.

Histogram A has a larger standard deviation; the values are spread widely around the mean.

Both plots show a normal distribution with a symmetrical spread of values about the mean.

Histogram B has a smaller standard deviation; the values are clustered more tightly around the mean.

Calculating s
Standard deviation is easily calculated using a spreadsheet.

$$s = \sqrt{\frac{\sum x^2 - ((\sum x)^2 / n)}{n}}$$

($\sum x$) = sum of value x
$\sum x^2$ = sum of value x^2
n = sample size

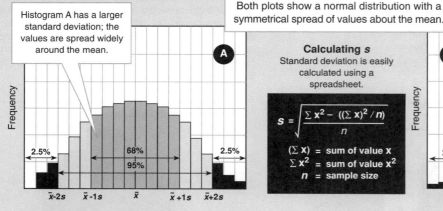

Case Study: Fern Reproduction

Raw data (below) and descriptive statistics (right) from a survey of the number of spores found on the fronds of a fern plant.

Fern spores

Raw data: Number of spores per frond

64	60	64	62	68	66	63
69	70	63	70	70	63	62
71	69	59	70	66	61	70
67	64	63	64			

$\dfrac{\text{Total of data entries}}{\text{Number of entries}} = \dfrac{1641}{25} = 66$ spores

Mean

Number of spores per frond (in rank order)	
59	66
60	66
61	67
62	68
62	69
63	69
63	70
63	70
63	70
64	70
64	70
64	71
64	

Median → 64

Spores per frond	Tally	Total
59	✓	1
60	✓	1
61	✓	1
62	✓✓	2
63	✓✓✓✓	4
64	✓✓✓✓	4
65		0
66	✓✓	2
67	✓	1
68	✓	1
69	✓✓	2
70	✓✓✓✓✓	5
71	✓	1

Mode → 70

1. Give a reason for the differences between the mean, median, and mode of the fern spore data:

   _____
   _____
   _____
   _____

2. Calculate the mean, median, and mode of the data on beetle masses below. Draw up a tally chart and show all calculations:

Beetle masses (g)		
2.2	2.1	2.6
2.5	2.4	2.8
2.5	2.7	2.5
2.6	2.6	2.5
2.2	2.8	2.4

The Reliability of the Mean

You have already seen how to use the **standard deviation** (s) to quantify the spread or **dispersion** in your data. The **variance** (s^2) is another such measure of dispersion, but the standard deviation is usually the preferred of these two measures because it is expressed in the original units. Usually, you will also want to know how good your sample mean ($\bar{x}$) is as an estimate of the true population mean (μ). This can be indicated by the standard error of the mean (or just **standard error** or SE). **SE** is often used as an error measurement simply because it is small, rather than for any good statistical reason. However, it is does allow you calculate the **95% confidence interval (95% CI)**. The calculation and use of 95% CIs is outlined below and the following page. By the end of this activity you should be able to:

- Enter data and calculate descriptive statistics using a spreadsheet program such as *Microsoft Excel*. You can follow this procedure for any set of data.
- Calculate standard error and 95% confidence intervals for sample data and plot these data appropriately with error bars.
- Interpret the graphically presented data and reach tentative conclusions about the findings of the experiment.

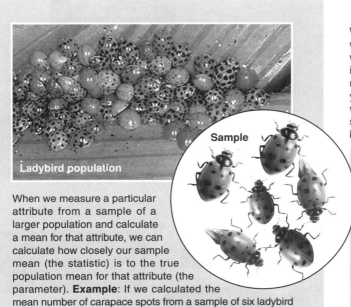

Ladybird population

When we measure a particular attribute from a sample of a larger population and calculate a mean for that attribute, we can calculate how closely our sample mean (the statistic) is to the true population mean for that attribute (the parameter). **Example**: If we calculated the mean number of carapace spots from a sample of six ladybird beetles, how reliable is this statistic as an indicator of the mean number of carapace spots in the whole population? We can find out by calculating the **95% confidence interval**.

Reliability of the Sample Mean

When we take measurements from samples of a larger population, we are using those samples as indicators of the trends in the whole population. Therefore, when we calculate a sample mean, it is useful to know how close that value is to the true population mean (μ). This is not merely an academic exercise; it will enable you to make **inferences** about the aspect of the population in which you are interested. For this reason, statistics based on samples and used to estimate population parameters are called **inferential statistics**.

The Standard Error (SE)

The standard error (SE) is simple to calculate and is usually a small value. Standard error is given by:

$$SE = \frac{s}{\sqrt{n}}$$

where s = the standard deviation, and n = sample size.

Standard errors are sometimes plotted as error bars on graphs, but it is more meaningful to plot the **95% confidence intervals** (see box below). All calculations are easily made using a spreadsheet (see opposite).

The 95% Confidence Interval

SE is required to calculate the 95% confidence interval (CI) of the mean. This is given by:

$$95\% \text{ CI} = SE \times t_{P(n-1)}$$

Do not be alarmed by this calculation; once you have calculated the value of the SE, it is a simple matter to multiply this value by the value of t at $P = 0.05$ (from the t table) for the appropriate degrees of freedom (df) for your sample ($n - 1$).

For example: where the SE = 0.6 and the sample size is 10, the calculation of the 95% CI is:

$$95\% \text{ CI} = 0.6 \times 2.262 = 1.36$$

Part of the t table is given to the right for $P = 0.05$. Note that, as the sample becomes very large, the value of t becomes smaller. For very large samples, t is fixed at 1.96, so the 95% CI is slightly less than twice the SE.

All these statistics, including a plot of the data with Y error bars, can be calculated using a program such as *Microsoft Excel* (opposite).

Critical values of Student's t distribution at $P = 0.05$.

df	P
	0.05
1	12.71
2	4.303
3	3.182
4	2.776
5	2.571
6	2.447
7	2.365
8	2.306
9	2.262
10	2.228
20	2.086
30	2.042
40	2.021
60	2.000
120	1.980
>120	1.960

Value of t at $n-1 = 9$

Maximum value of t at this level of P

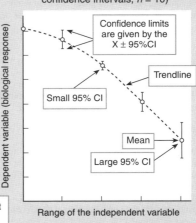

Relationship of Y against X (± 95% confidence intervals, $n = 10$)

Plotting your confidence intervals

Once you have calculated the 95% CI for the means in your data set, you can plot them as error bars on your graph. Note that the **95% confidence limits** are given by the value of the **mean ± 95%CI**. A 95% confidence limit (i.e. $P = 0.05$) tells you that, on average, 95 times out of 100, the limits will contain the true population mean.

© Biozone International 2006-2007
Photocopying Prohibited

Related activities: Descriptive Statistics, Taking the Next Step

Comparing Treatments Using Descriptive Statistics

In an experiment, the growth of newborn rats on four different feeds was compared by weighing young rats after 28 days on each of four feeding regimes. The suitability of each food type for maximizing growth in the first month of life was evaluated by comparing the means of the four experimental groups. Each group comprised 10 individual rats. All 40 newborns were born to sibling mothers with the same feeding history. For this activity, follow the steps outlined below and reproduce them yourself.

Calculating Descriptive Statistics

Entering your data and calculating descriptive statistics.

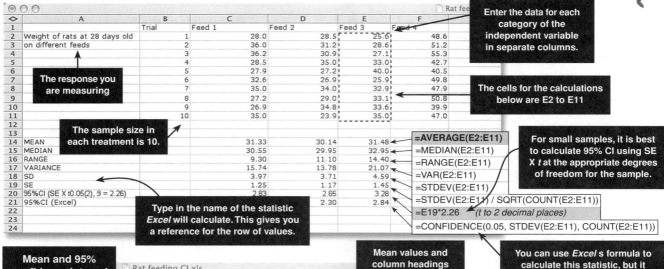

Drawing the Graph

To plot the graph, you will need to enter the data values you want to plot in a format that *Excel* can use (above). To do this, enter the values in columns under each category.

- Each column will have two entries: mean and 95% CI. In this case, we want to plot the mean weight of 28 day rats fed on different foods and add the 95% confidence intervals as error bars.

- The independent variable is categorical, so the correct graph type is a column chart. Select the row of mean values (including column headings)

1 From the menu bar choose: **Insert** > **Chart** > **Column**. This is **Step 1** in the Chart Wizard. Click **Next**.

2 At **Step 2**, click **Next**.

3 At **Step 3**, you have the option to add a title, labels for your X and Y axes, turn off gridlines, and add (or remove) a legend. When you have added all the information you want, click **Next**.

4 At **Step 4**, specify the chart location. It should appear "as object in" Sheet 1 by default. Click on the chart and move it to reveal the data.

5 A chart will appear on the screen. **Right click** (Ctrl-click on Mac) on any part of any column and choose **Format data series**. To add error bars, select the **Y error bars** tab, and click on the symbol that shows Display both. Click on Custom, and use the data selection window to select the row of 95% CI data for "+" and "–" fields.

6 Click on OK and your chart will plot with error bars.

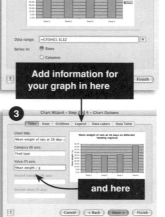

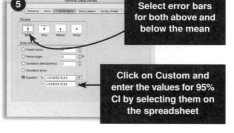

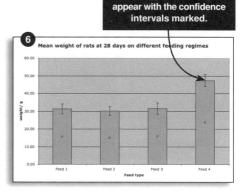

© Biozone International 2006-2007
Photocopying Prohibited

Linear Regression

Regression is a test for an association, relationship, or trend between two variables. It is suitable for continuous data when you have a reason to believe that the changes in one variable cause changes in the other, i.e. regression assumes a cause and effect. A regression is also predictive; the **regression equation** will be able to predict unknown values of the Y variable within the range covered by the data. Linear regression is the simplest functional relationship of one variable to another. If your data are appropriate for this analysis, they will plot as a straight line spread on a scatter graph. It is best to perform your regression on the raw data, because information is lost when the calculation is performed on mean values. If your data plot is not linear, you have the choice of plotting a non-linear regression (see the next activity) or transforming your data to make them linear.

Linear regression is a simple relationship where the change in the independent variable causes a corresponding change in the dependent variable in a simple linear fashion. A line is fitted to the data and gives the values of the slope and intercept of the line (the computer does this for you).

Linear regressions are simple to perform using a computer program such as *Microsoft Excel*. The steps for doing this are outlined here.

Clutch Size vs Body Size in *Daphnia*

Daphnia is a small, freshwater crustacean. In *Daphnia*, body size is a large determinant of the number of eggs they can carry (this female has none). The relationship between body size and clutch size can be described with a regression.

1 Clutch size (number of eggs per female) was estimated for 50 females, and body length was measured to the nearest 0.01 mm for the same individuals to give 50 paired values. These data values were entered directly into *Microsoft Excel*.

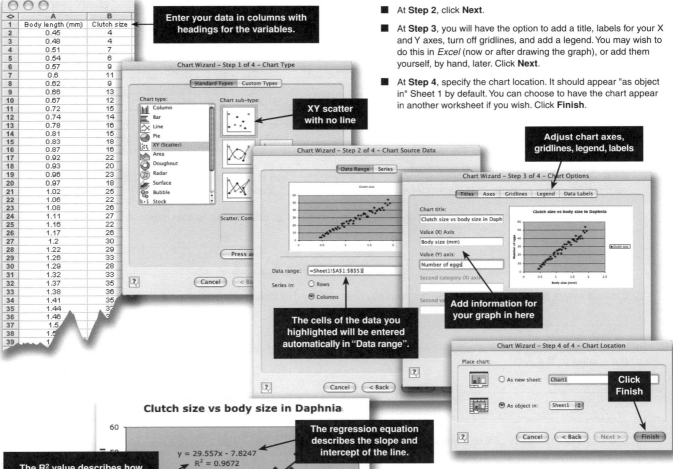

2 To draw the graph, highlight the data columns: "Body length" and "Clutch size".

- From the menu bar choose **Insert** > **Chart** > **XY (Scatter)**, and click on the option with no line.
  This is **Step 1** of 4 in the Chart Wizard. Click **Next**.

- At **Step 2**, click **Next**.

- At **Step 3**, you will have the option to add a title, labels for your X and Y axes, turn off gridlines, and add a legend. You may wish to do this in *Excel* (now or after drawing the graph), or add them yourself, by hand, later. Click **Next**.

- At **Step 4**, specify the chart location. It should appear "as object in" Sheet 1 by default. You can choose to have the chart appear in another worksheet if you wish. Click **Finish**.

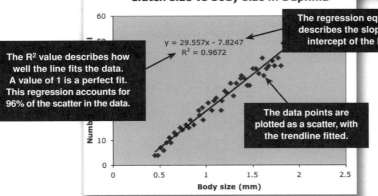

3 A chart will appear on the screen. If necessary, click on the chart and move it slightly to reveal the data. To add the trendline, **right click** (Ctrl-click on Mac) on any data point on the graph, choose **Add Trendline** and choose **Type** > **Linear**. Click on the Options tab and click on the check boxes for **Display equation on chart** and **Display R^2 value on chart**. Click **OK**. The equation is a text box that can be moved by clicking and dragging with the mouse.

Labels for the X and Y axes are added by right clicking within the plot area, and choosing Chart Options and the Titles tab.

Related activities: Transforming Raw Data, Drawing Scatter Plots, Descriptive Statistics, Non-Linear Regression

1. Explain why the data for clutch size and body size in *Daphnia* are appropriate for analysis with a linear regression:

_____

_____

_____

_____

2. A student's experiment, investigated the effect of increasing seawater dilution on cumulative weight gain (as an indication of osmoregulatory ability) of a common shore crab. Six crabs in total were used in the experiment. Three were placed in seawater dilution of 75:25 (75% seawater) and three were placed in a seawater dilution of 50:50 (50% seawater). Cumulative weight gain in each of the six crabs was measured at regular intervals over a period of 30 minutes. The results are tabulated below:

Time (min)	Crab weight gain at 75% seawater (mg)			Crab weight gain at 50% seawater (mg)		
	Crab number			Crab number		
	1	2	3	4	5	6
3	3.80	4.00	4.00	5.60	6.20	5.80
6	8.00	8.30	7.70	11.20	11.60	11.90
9	11.50	11.00	9.50	17.00	17.60	17.20
12	14.80	15.10	15.20	23.50	23.60	24.00
15	18.90	19.50	19.70	29.00	28.20	28.60
18	23.50	22.90	23.80	33.00	32.50	32.70
21	26.50	26.90	26.70	37.50	37.60	39.00
24	31.50	32.00	31.20	43.10	43.50	43.60
27	35.00	35.50	35.50	48.00	48.10	47.50
30	40.00	40.10	41.20	53.00	52.60	52.80

(a) Following the steps outlined on the opposite page, enter these data in an appropriate way on a spreadsheet (e.g. *Microsoft Excel*) and plot a scatter plot to show the relationship between time and weight gain in crabs held at two different dilutions of seawater. Be sure to add appropriate titles and axis labels to your graph as you proceed.

(b) Fit a trendline to your plot, and display the regression equation and the R^2 value. When you have finished your analysis, staple a printout of the completed spreadsheet into your workbook.

(c) Describe the relationship between time and weight gain in crabs at different seawater dilutions: _____

_____

_____

(d) Explain why a linear regression is an appropriate analysis for this data set: _____

_____

(e) Make a statement about the results of your regression analysis with respect to how well the regression accounts for the scatter in the data:

_____

_____

(f) Discuss the limitations of this experimental design: _____

_____

_____

_____

(g) Suggest ways in which the experimental design could be improved: _____

_____

_____

Non-linear Regression

The degree to which organisms respond physiologically to their environment depends on the type of organism and how tolerant they are of environmental changes. Biological responses are often not described by a simple linear relationship (although they may be linear when transformed). Many metabolic relationships are non-linear because organisms function most efficiently within certain environmental limits. When you plot the data for these kinds of relationships, they will plot out as something other than a straight line. An example of non-linear regression is described below for egg development time in *Daphnia*.

Egg Development Time in *Daphnia*

In *Daphnia*, the **egg development time** (EDT) is the time from egg deposition to release of the young from the brood pouch. From an experiment that measured EDT at controlled temperatures, means and 95% confidence intervals were calculated from a sample size of five animals at each of five temperatures between 9 and 25°C. The aim of this activity is to demonstrate an analysis of a non-linear metabolic relationship by:

- Plotting egg development time (in days) against water temperature (in °C).
- Adding error bars for the means.
- Fitting a nonlinear regression and showing the line equation and the R^2 value, which is a measure of appropriateness of the model describing the line. Here, we have used a power model, but if you have a non-linear relationship, you could also try a polynomial as a model (another choice in Chart Wizard).

1 Create a table in your spreadsheet with your summary data in columns, as shown below. Your raw data may already be entered in the spreadsheet and you can use the spreadsheet to calculate the **summary statistics**. See the activity on the Analysis of Variance for guidance on how to do this.

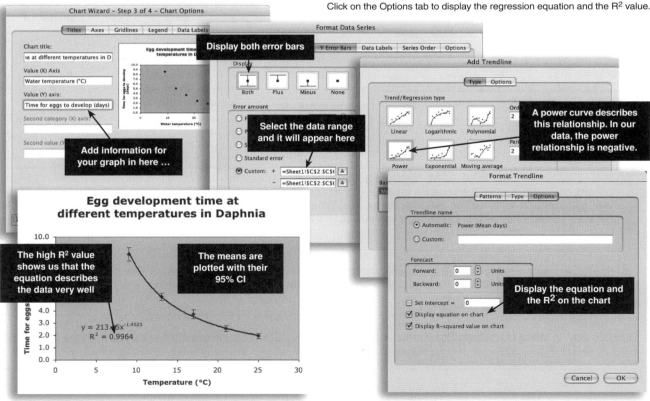

2 To perform this regression, follow the same sequence of steps to those in the earlier activity (linear regression).

- First, highlight the data columns: **Temperature** and **Mean days**. From the menu bar choose **Insert > Chart > XY (Scatter)**, and click on the option with no line. This is **Step 1** of 4 in the Chart Wizard. Click **Next**.
- At **Step 2**, click **Next**.
- At **Step 3**, you have the option to add a title, labels for your X and Y axes, turn off gridlines, and add (or remove) a legend. When you have added all the information you want, click **Next**.
- At **Step 4**, specify the chart location. It should appear "as object in" Sheet 1 by default. Click on the chart and move it to reveal the data.

3 A chart will appear on the screen. **Right click** (Ctrl-click on Mac) on any data point on the graph and choose **Format data series**. To add error bars, select the **Y error bars** tab, and click on the symbol that shows Display both. Click on Custom, and use the data selection window to select the data under the 95% CI column label. To add **trendline**, **right click** (Ctrl-click on Mac) on any data point on the graph, and choose **Add Trendline** and **Power**. Click on the Options tab to display the regression equation and the R^2 value.

1. Describe the relationship between temperature and EDT in *Daphnia*: _____

The Student's t Test

The Student's *t* test is a commonly used test when comparing two sample means, e.g. means for a treatment and a control in an experiment, or the means of some measured characteristic between two animal or plant populations. The test is a powerful one, i.e. it is a good test for distinguishing real but marginal differences between samples. The *t* test is a simple test to apply, but it is only valid for certain situations. It is a two-group test and is not appropriate for multiple use i.e. sample 1 vs 2, then sample 1 vs 3. *You must have only two sample means to compare.* You are also assuming that the data have a normal (not skewed) distribution, and the scatter (standard deviations) of the data points is similar for both samples. You may wish to exclude obvious outliers from your data set for this reason. Below is a simple example outlining the general steps involved in the Student's *t* test. The following is a simple example using a set of data from a fictitious experiment involving a treatment and a control (the units are not relevant in this case, only the values). A portion of the Student's *t* table is provided, sufficient to carry out the test. Follow the example through, making sure that you understand what is being done at each step.

Steps in performing a Student's *t* test	Explanatory notes
Step 1 *Calculate basic summary statistics for your two data sets* Control (A): 6.6, 5.5, 6.8, 5.8, 6.1, 5.9 $n_A = 6$, $\bar{x}_A = 6.12$, $s_A = 0.496$ Treatment (B): 6.3, 7.2, 6.5, 7.1, 7.5, 7.3 $n_B = 6$, $\bar{x}_B = 6.98$, $s_B = 0.475$	n_A and n_B are the number of values in the first and second data sets respectively (these need not be the same). $\bar{x}$ is the mean. s is the standard deviation (a measure of scatter in the data).
Step 2 *Set up and state your null hypothesis (H_0)* H_0: there is no treatment effect. The differences in the data sets are the result of chance variation only and they are not really different	The alternative hypothesis is that there is a treatment effect and the two sets of data are truly different.
Step 3 *Decide if your test is one or two tailed* This tells you what section of the *t* table to consult. Most biological tests are two-tailed. Very few are one-tailed.	A one-tailed test looks for a difference only in one particular direction. A two-tailed test looks for any difference (+ or –).
Step 4 *Calculate the t statistic* For our sample data above the calculated value of *t* is –3.09. The degrees of freedom (df) are $n_1 + n_2 - 2 = 10$. Calculation of the *t* value uses the variance which is simply the square of the standard deviation (s^2). You may compute the *t* value by entering your data onto a computer and using a simple statistical program.	It does not matter if your calculated *t* value is a positive or negative (the sign is irrelevant). If you do not have access to a statistical program, computation of *t* is not difficult. Step 4 (calculating *t*) is described in the following *t* test exercises.
Step 5 *Consult the t table of critical values* Selected critical values for Student's *t* statistic (two-tailed test) 	Degrees of freedom	$P = 0.05$	$P = 0.01$	$P = 0.001$
---	---	---	---
5	2.57	4.03	6.87
10	(2.23)	3.17	4.59
15	2.13	2.95	4.07
20	2.09	2.85	3.85 | Critical value of *t* for 10 degrees of freedom. The calculated *t* value must exceed this	The absolute value of the *t* statistic (3.09) well exceeds the critical value for $P = 0.05$ at 10 degrees of freedom. *We can reject H_0 and conclude that the means are different at the 5% level of significance.* If the calculated absolute value of *t* had been less than 2.23, we could not have rejected H_0.

1. (a) In an experiment, data values were obtained from four plants in experimental conditions and three plants in control conditions. The mean values for each data set (control and experimental conditions) were calculated. The *t* value was calculated to be 2.16. The null hypothesis was: "The plants in the control and experimental conditions are not different". State whether the calculated *t* value supports the null hypothesis or its alternative (consult *t* table above):

   _____

   (b) The experiment was repeated, but this time using 6 control and 6 "experimental" plants. The new *t* value was 2.54. State whether the calculated *t* value supports the null hypothesis or its alternative now:

   _____

2. Explain why, in terms of applying Student's *t* test, extreme data values (outliers) are often excluded from the data set(s):

   _____
   _____

3. Explain what you understand by statistical significance (for any statistical test): _____
   _____
   _____

Student's t Test Exercise

Data from two flour beetle populations are provided below. The numbers of beetles in each of ten samples were counted. The experimenter wanted to test if the densities of the two populations were significantly different. The exercise below involves manual computation to determine a *t* value. Follow the steps to complete the test. You can also use a spreadsheet program such as *Microsoft Excel* to do the computations (overleaf) or perform the entire analysis for you (see the Teacher Resource CD-ROM).

1. (a) Complete the calculations to perform the *t* test for these two populations. Some calculations are provided for you.

x (counts)		x − x̄ (deviation from the mean)		(x − x̄)² (deviation from mean)²	
Popn A	Popn B	Popn A	Popn B	Popn A	Popn B
465	310	9.3	−10.6	86.5	112.4
475	310	19.3	−10.6	372.5	112.4
415	290				
480	355				
436	350				
435	335				
445	295				
460	315				
471	316				
475	330				
$n_A = 10$	$n_B = 10$	The sum of each column is called the sum of squares		$\Sigma(x-\bar{x})^2$	$\Sigma(x-\bar{x})^2$

The number of samples in each data set

Step 1: Summary statistics
Tabulate the data as shown in the first 2 columns of the table (left). Calculate the mean and give the *n* value for each data set. Compute the standard deviation if you wish.

Popn A $\bar{x}_A = 455.7$ Popn B $\bar{x}_B = 320.6$
         $n_A = 10$ $n_B = 10$
         $s_A = 21.76$ $s_B = 21.64$

Step 2: State your null hypothesis

Step 3: Decide if your test is one or two-tailed

Calculating the t value

Step 4a: Calculate sums of squares
Complete the computations outlined in the table left. The sum of each of the final two columns (left) is called the sum of squares.

(b) The variance for population A: $s^2_A =$

The variance for population B: $s^2_B =$

Step 4b: Calculate the variances
Calculate the variance (s^2) for each set of data. This is the sum of squares divided by $n-1$ (number of samples in each data set − 1). In this case the *n* values are the same, but they need not be.

$$s^2_A = \frac{\Sigma(x-\bar{x})^2_{(A)}}{n_A - 1} \qquad s^2_B = \frac{\Sigma(x-\bar{x})^2_{(B)}}{n_B - 1}$$

(c) The difference between the population means

$(\bar{x}_A - \bar{x}_B) =$

Step 4c: Difference between means
Calculate the *actual* difference between the means
$(\bar{x}_A - \bar{x}_B)$

(d) *t* (calculated) =

Step 4d: Calculate t
Calculate the *t* value. Ask for assistance if you find interpreting the lower part of the equation difficult

$$t = \frac{(\bar{x}_A - \bar{x}_B)}{\sqrt{\dfrac{s^2_A}{n_A} + \dfrac{s^2_B}{n_B}}}$$

(e) Determine degrees of freedom (d.f.)

d.f. ($n_A + n_B - 2$) =

Step 4e: Determine degrees of freedom
Degrees of freedom (d.f.) are defined by the number of samples (e.g. counts) taken: d.f. = $n_A + n_B - 2$ where n_A and n_B are the number of counts in each of populations A and B.

(f) *P* =

t (critical value) =

(g) Your decision is:

Step 5: Consult the t table
Consult the *t*-tables (opposite page) for the critical *t* value at the appropriate degrees of freedom and the acceptable probability level (e.g. $P = 0.05$).

Step 5a: Make your decision
Make your decision whether or not to reject H_0. If your *t* value is large enough you may be able to reject H_0 at a lower *P* value (e.g. 0.001), increasing your confidence in the alternative hypothesis.

2. The previous example (manual calculation for two beetle populations) is outlined below in a spreadsheet (created in *Microsoft Excel*). The spreadsheet has been shown in a special mode with the formulae displayed. Normally, when using a spreadsheet, the calculated values will appear as the calculation is completed (entered) and a formula is visible only when you click into an individual cell. When setting up a spreadsheet, you can arrange your calculating cells wherever you wish. What is important is that you accurately identify the cells being used for each calculation. Also provided below is a summary of the spreadsheet notations used and a table of critical values of t at different levels of P. Note that, for brevity, only some probability values have been shown. To be significant at the appropriate level of probability, calculated values must be greater than those in the table for the appropriate degrees of freedom.

(a) Using the data in question 1, set up a spreadsheet as indicated below to calculate t. Save your spreadsheet. Print it out and staple the print-out into your workbook.

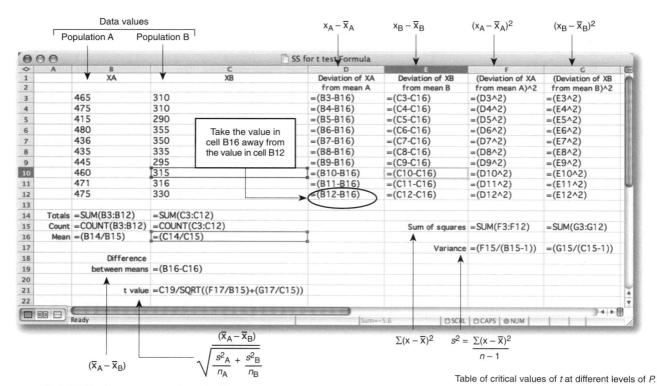

Notation	Meaning
Columns and rows	Columns are denoted A, B, C ... at the top of the spreadsheet, rows are 1, 2, 3, ... on the left. Using this notation a cell can be located e.g. C3
=	An "equals" sign *before* other entries in a cell denotes a formula.
( )	Parentheses are used to group together terms for a single calculation. This is important for larger calculations (see cell C21 above)
C3:C12	Cell locations are separated by a colon. C3:C12 means "every cell between and including C3 and C12"
SUM	Denotes that what follows is added up. =SUM(C3:C12) means "add up the values in cells C3 down to C12"
COUNT	Denotes that the number of values is counted =COUNT(C3:C12) means "count up the number of values in cells C3 down to C12"
SQRT	Denotes "take the square root of what follows"
^2	Denotes an exponent e.g. x^2 means that value x is squared.

Above is a table explaining some of the spreadsheet notations used for the calculation of the t value for the exercise on the previous page. It is not meant to be an exhaustive list for all spreadsheet work, but it should help you to become familiar with some of the terms and how they are used. This list applies to *Microsoft Excel*. Different spreadsheets may use different notations. These will be described in the spreadsheet manual.

Table of critical values of t at different levels of P.

Degrees of freedom	Level of Probability		
	0.05	0.01	0.001
1	12.71	63.66	636.6
2	4.303	9.925	31.60
3	3.182	5.841	12.92
4	2.776	4.604	8.610
5	2.571	4.032	6.869
6	2.447	3.707	5.959
7	2.365	3.499	5.408
8	2.306	3.355	5.041
9	2.262	3.250	4.781
10	2.228	3.169	4.587
11	2.201	3.106	4.437
12	2.179	3.055	4.318
13	2.160	3.012	4.221
14	2.145	2.977	4.140
15	2.131	2.947	4.073
16	2.120	2.921	4.015
17	2.110	2.898	3.965
18	2.101	2.878	3.922
19	2.093	2.861	3.883
20	2.086	2.845	3.850

(b) Save your spreadsheet under a different name and enter the following new data values for population B: **425, 478, 428, 465, 439, 475, 469, 445, 421, 438**. Notice that, as you enter the new values, the calculations are updated over the entire spreadsheet. Re-run the t-test using the new t value. State your decision for the two populations now:

New t value: _____ Decision on null hypothesis (delete one): Reject / Do not reject

Comparing More Than Two Groups

The Student's *t* test is limited to comparing two groups of data. To compare more than two groups at once you need to use a test that is appropriate to this aim. One such test, appropriate for normally distributed data, is an analysis of variance (**ANOVA**), as described in the next activity. A good place to start with any such an analysis is to plot your data, together with some measure of the reliability of the statistic you calculate as an indicator of the true population parameter. For normally distributed data, this is likely to be the mean and the 95% confidence interval (95% CI). See *The Reliability of the Mean* for an introduction to this statistic. In the example described below, students recorded the survival of weevil larvae on five different pasture types and calculated descriptive statistics for the data. After you have worked through the analysis, you should be able to enter your own data and calculate descriptive statistics using *Microsoft Excel*. The plot and full analysis of these data are presented in the next activity.

Comparing the Means of More Than Two Experimental Groups

Research has indicated that different pastures have different susceptibility to infestation by a pest insect, the clover root weevil (left). Armed with this knowledge, two students decided to investigate the effect of pasture type on the survival of clover root weevils. The students chose five pasture types, and recorded the number of weevil larvae (from a total of 50) surviving in each pasture type after a period of 14 days. Six experimental pots were set up for each pasture type ($n = 6$). Their results and the first part of their analysis (calculating the descriptive statistics) are presented in this activity.

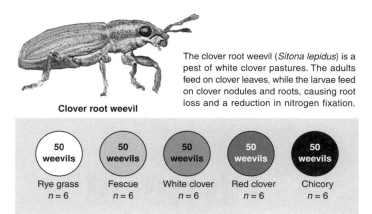

The clover root weevil (*Sitona lepidus*) is a pest of white clover pastures. The adults feed on clover leaves, while the larvae feed on clover nodules and roots, causing root loss and a reduction in nitrogen fixation.

1 Calculating Descriptive Statistics

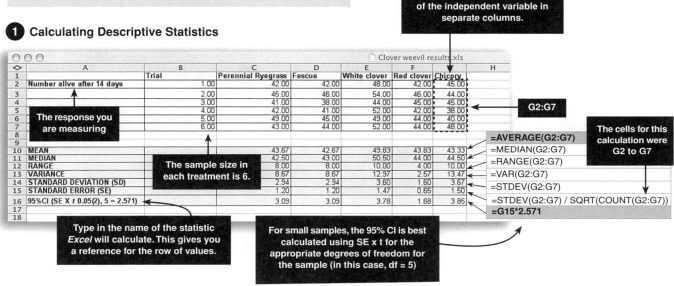

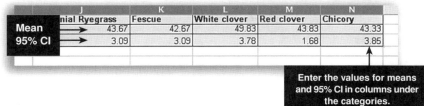

2 To plot a graph in *Excel*, you will need to enter the values for the data you want to plot in a format that *Excel* can use. To do this, enter the values in columns under each category (left). Each column will have two entries: mean and 95% CI. In this case, we want to plot the mean survival of weevils in different pasture types and add the 95% CI as error bars. You are now ready to plot the data set.

1. Identify the number of treatments in this experimental design: _____

2. Identify the independent variable for this experiment: _____

3. Identify the dependent variable for this experiment: _____

4. Identify the type of graph you would use to plot this data set and explain your choice: _____

5. Explain what the 95% confidence interval tells you about the means for each treatment: _____

Analysis of Variance

Analysis of variance (or **ANOVA**) could be considered beyond the scope of the statistical analyses you would do at school. However, using *Excel* (previous activity and below), it is not difficult and is an appropriate test for a number of situations in experimental biology. It is described below for the data on weevil survival described in the previous activity. A plot of the data is also described as an aid to interpreting the statistical analysis. An ANOVA may also be appropriate when the independent variable is continuous, but in these instances a regression is likely to be a more suitable analysis (*see pages 36-38*). After you have worked through this activity you should be able to:

- Use *Excel* to plot data appropriately with error bars.
- Interpret the graphically presented data and reach tentative conclusions about the findings of the experiment.
- Support your conclusions with an ANOVA, in *Microsoft Excel*, to test the significance of differences in the sample data.

Drawing the Graph

Recall the design of this experiment. The independent variable is categorical, so the correct graph type is a column chart. To plot the graph, select the row of mean values (including column headings) from the small table of results you constructed (screen, right).

1. From the menu bar choose: **Insert > Chart > Column**. This is **STEP 1** in the Chart Wizard. Click **Next**.

2. At **STEP 2** in Chart Wizard: The data range you have selected (the source data for the chart) will appear in the "Data range" window. Click **Next**.

3. At **STEP 3** in Chart Wizard: You have the option to add a title, labels for your X and Y axes, turn off gridlines, and add (or remove) a legend (a legend is useful when you have two or more columns, such as males and females, for each treatment). When you have added all the information you want, click **Next**.

4. At **STEP 4** in Chart Wizard: Specify the chart location. It should appear in the active sheet by default. Click on the chart and move it to reveal the data.

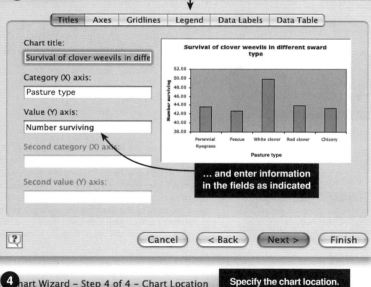

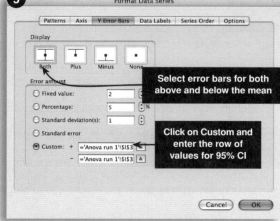

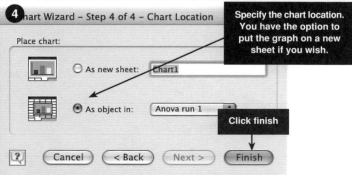

5. **Adding error bars:** A chart will appear. **Right click** (Ctrl-click on Mac) on any part of any column and choose **Format data series** (below). To add error bars, select the **Y error bars** tab, and click on the symbol that shows **Display both**. Click on **Custom**, and use the data selection window to select the row of 95% CI data. Click OK.

Related activities: Comparing More Than Two Groups

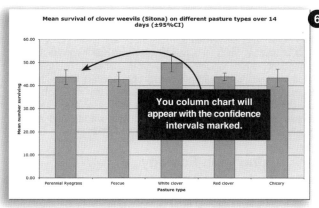

After you have added error bars using the "Format data series" option, your chart will plot with error bars. These error bars represent the 95% confidence intervals for the mean of each treatment. The completed plot gives you a visual representation of your data and an immediate idea as to how confident you can be about the differences between the five treatments. You are now ready to run your analysis of variance.

About ANOVA

ANOVA is a test for normally distributed data, which can determine whether or not there is a real (statistically significant) difference between two or more sample means (an ANOVA for two means is a *t* test). ANOVA does this by accounting for the amount of variability (in the measured variable) within a group and comparing it to the amount of variability between the groups (treatments). It is an ideal test when you are looking for the effect of a particular treatment on some kind of biological response, e.g. plant growth with different fertilisers. ANOVA is a very useful test for comparing multiple samples because, like the *t* test, it does not require large sample sizes.

Running the ANOVA

ANOVA is part of the *Excel* "Data Analysis Toolpak", which is part of normal *Excel*, but is not always installed. If there is no Data Analysis option on the Tools menu, you need to run *Excel* from the original installation disk to install the "Analysis Toolpak add-in".

From the Tools menu select Data Analysis and **ANOVA single factor**. This brings up the ANOVA dialogue box (numbered points 1 and 2). Here we are concerned with the effect of one variable on another, so the ANOVA is single factor (also called one way ANOVA).

Click in the **Output Range** box and click on a free cell on the worksheet which will become the left cell of the results table. Click OK.

The output is a large data table (number 3 below) and you will need to adjust the number of significant figures to 2 or 3 (You do this from the Format menu > Cells > Number). The most important cell is the *P* value, which, as usual is the probability that the null hypothesis (of no difference between the groups) is true.

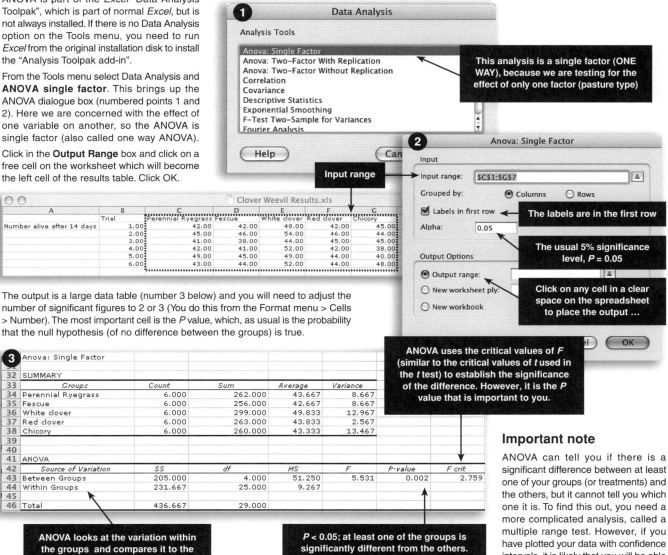

Important note

ANOVA can tell you if there is a significant difference between at least one of your groups (or treatments) and the others, but it cannot tell you which one it is. To find this out, you need a more complicated analysis, called a multiple range test. However, if you have plotted your data with confidence intervals, it is likely that you will be able to find this out from your graph.

1. The ANOVA cannot tell us which one (or more) of the groups is significantly different from the others. Explain how you could you use your graphical analysis to answer this question:

Using the Chi-Squared Test in Ecology

The **chi-squared test** (χ^2), like the student's *t* test, is a test for difference between data sets, but it is used when you are working with frequencies (counts) rather than measurements. It is a simple test to perform but the data must meet the requirements of the test. These are as follows:
- It can only be used for data that are raw counts (not measurements or derived data such as percentages).
- It is used to compare an experimental result with an expected theoretical outcome (e.g. an expected Mendelian ratio or a theoretical value indicating "no preference" or "no difference" between groups in some sort of response such as habitat or microclimate preference).
- It is not a valid test when sample sizes are small (<20).

Like all statistical tests, it aims to test the null hypothesis; the hypothesis of no difference between groups of data. The following exercise is a worked example using chi-squared for testing an ecological study of habitat preference. As with most of these simple statistical tests, chi-squared is easily calculated using a spreadsheet. Guidelines for this as well as an additional activity are available on the Teacher Resource CD-ROM.

Using χ^2 in Ecology

In an investigation of the ecological niche of the mangrove, *Avicennia marina var. resinifera*, the density of pneumatophores was measured in regions with different substrate. The mangrove trees were selected from four different areas: mostly sand, some sand, mostly mud, and some mud. Note that the variable, substrate type, is categorical in this case. Quadrats (1 m by 1 m) were placed around a large number of trees in each of these four areas and the numbers of pneumatophores were counted. Chi-squared was used to compare the observed results for pneumatophore density (as follows) to an expected outcome of no difference in density between substrates.

Pneumatophores

Mangrove pneumatophore density in different substrate areas				
Mostly sand	85	Mostly mud	130
Some sand	102	Some mud	123

Using χ^2, the probability of this result being consistent with the expected result could be tested. Worked example as follows:

Step 1: Calculate the expected value (E)
In this case, this is the sum of the observed values divided by the number of categories. $\frac{440}{4} = 110$

Step 2: Calculate O – E
The difference between the observed and expected values is calculated as a measure of the deviation from a predicted result. Since some deviations are negative, they are all squared to give positive values. This step is usually performed as part of a tabulation (right, darker gray column).

Step 3: Calculate the value of χ^2
$$\chi^2 = \sum \frac{(O-E)^2}{E}$$

Where: O = the observed result
E = the expected result
Σ = sum of

The calculated χ^2 value is given at the bottom right of the last column in the tabulation.

Category	O	E	O – E	(O – E)²	$\frac{(O-E)^2}{E}$
Mostly sand	85	110	–25	625	5.68
Some sand	102	110	–8	64	0.58
Mostly mud	130	110	20	400	3.64
Some mud	123	110	13	169	1.54

Total = 440 χ^2 $\Sigma = 11.44$

Step 5a: Using the χ^2 table
On the χ^2 table (part reproduced in Table 1 below) with 3 degrees of freedom, the calculated value for χ^2 of 11.44 corresponds to a probability of between 0.01 and 0.001 (see arrow). *This means that by chance alone a χ^2 value of 11.44 could be expected between 1% and 0.1% of the time.*

Step 4: Calculating degrees of freedom
The probability that any particular χ^2 value could be exceeded by chance depends on the number of degrees of freedom. This is simply **one less than the total number of categories** (this is the number that could vary independently without affecting the last value). *In this case: 4–1 = 3.*

Step 5b: Using the χ^2 table
The probability of between 0.1 and 0.01 is lower than the 0.05 value which is generally regarded as significant. The null hypothesis can be rejected and we have reason to believe that the observed results differ significantly from the expected (at $P = 0.05$).

Table 1: Critical values of χ^2 at different levels of probability. By convention, the critical probability for rejecting the null hypothesis (H_0) is 5%. If the test statistic is less than the tabulated critical value for $P = 0.05$ we cannot reject H_0 and the result is not significant. If the test statistic is greater than the tabulated value for $P = 0.05$ we reject H_0 in favour of the alternative hypothesis.

Degrees of freedom	Level of probability (P)									
	0.98	0.95	0.80	0.50	0.20	0.10	0.05	0.02	0.01	0.001
1	0.001	0.004	0.064	0.455	1.64	2.71	3.84	5.41	6.64	10.83
2	0.040	0.103	0.466	1.386	3.22	4.61	5.99	7.82	9.21	13.82
3	0.185	0.352	1.005	2.366	4.64	6.25	7.82	9.84	11.35	16.27
4	0.429	0.711	1.649	3.357	5.99	7.78	9.49	11.67	13.28	18.47
5	0.752	0.145	2.343	4.351	7.29	9.24	11.07	13.39	15.09	20.52

← Do not reject H_0 | Reject H_0 →

Chi-Squared Exercise in Ecology

The following exercise illustrates the use of chi-squared (χ^2) in ecological studies of habitat preference. In the first example, it is used for determining if the flat periwinkle *(Littorina littoralis)* shows significant preference for any of the four species of seaweeds with which it is found. Using quadrats, the numbers of periwinkles associated with each seaweed species were recorded. The data from this investigation are provided for you in Table 1. In the second example, the results of an investigation into habitat preference in woodlice (also called pillbugs, sowbugs, or slaters) are presented for analysis (Table 2).

1. (a) State your null hypothesis for this investigation (H_0):

    _____

    _____

    _____

   (b) State the alternative hypothesis (H_A): _____

    _____

    _____

Table 1: Number of periwinkles associated with different seaweed species

Seaweed species	Number of periwinkles
Spiral wrack	9
Bladder wrack	28
Toothed wrack	19
Knotted wrack	64

2. Use the chi-squared test to determine if the differences observed between the samples are significant or if they can be attributed to chance alone. The table of critical values of χ^2 is provided in "The Chi-Squared Test" in *Skills in Biology*.

   (a) Enter the observed values (no. of periwinkles) and complete the table to calculate the χ^2 value:

   (b) Calculate χ^2 value using the equation:

   $$\chi^2 = \sum \frac{(O - E)^2}{E}$$

   $\chi^2 =$ _____

   (c) Calculate the degrees of freedom: _____

   (d) Using the χ^2 table, state the *P* value corresponding to your calculated χ^2 value:

   _____

   (e) State whether or not you reject your null hypothesis:

   reject H_0 / do not reject H_0 *(circle one)*

Category	O	E	O – E	(O – E)²	$\frac{(O-E)^2}{E}$
Spiral wrack					
Bladder wrack					
Toothed wrack					
Knotted wrack					
Σ					Σ

3. Students carried out an investigation into habitat preference in woodlice. In particular, they were wanting to know if the woodlice preferred a humid atmosphere to a dry one, as this may play a part in their choice of habitat. They designed a simple investigation to test this idea. The woodlice were randomly placed into a choice chamber for 5 minutes where they could choose between dry and humid conditions (atmosphere). The investigation consisted of five trials with ten woodlice used in each trial. Their results are shown on Table 2 (right):

   (a) State the null and alternative hypotheses (H_0 and H_A):

    _____

    _____

Table 2: Habitat preference in woodlice

Trial	Atmosphere	
	Dry	Humid
1	2	8
2	3	7
3	4	6
4	1	9
5	5	5

   Use a separate piece of paper (or a spreadsheet) to calculate the chi-squared value and summarize your answers below:

   (b) Calculate the χ^2 value: _____

   (c) Calculate the degrees of freedom and state the *P* value corresponding to your calculated χ^2 value: _____

   (d) State whether or not you reject your null hypothesis: reject H_0 / do not reject H_0 *(circle one)*

© Biozone International 2006-2007
Photocopying Prohibited

Related activities: Using the Chi-Squared Test in Ecology, Using the Chi-Squared Test in Genetics, Using Chi-Squared in Genetics

RDA 3

Using the Chi-Squared Test in Genetics

The **chi-squared test**, χ^2, is frequently used for testing the outcome of dihybrid crosses against an expected (predicted) Mendelian ratio, and it is appropriate for use in this way. When using the chi-squared test for this purpose, the null hypothesis predicts the ratio of offspring of different phenotypes according to the expected Mendelian ratio for the cross, assuming independent assortment of alleles (no linkage). Significant departures from the predicted Mendelian ratio indicate linkage of the alleles in question. Raw counts should be used and a large sample size is required for the test to be valid.

Using χ^2 in Mendelian Genetics

In a *Drosophila* genetics experiment, two individuals were crossed (the details of the cross are not relevant here). The predicted Mendelian ratios for the offspring of this cross were 1:1:1:1 for each of the four following phenotypes: gray body-long wing, gray body-vestigial wing, ebony body-long wing, ebony body-vestigial wing. The observed results of the cross were not exactly as predicted. The following numbers for each phenotype were observed in the offspring of the cross:

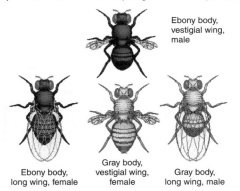

Ebony body, vestigial wing, male
Ebony body, long wing, female
Gray body, vestigial wing, female
Gray body, long wing, male

Images of *Drosophila* courtesy of **Newbyte Educational Software**: *Drosophila* Genetics Lab (www.newbyte.com)

Observed results of the example *Drosophila* cross			
Gray body, long wing	98	Ebony body, long wing	102
Gray body, vestigial wing	88	Ebony body, vestigial wing	112

Using χ^2, the probability of this result being consistent with a 1:1:1:1 ratio could be tested. Worked example as follows:

Step 1: Calculate the expected value (E)

In this case, this is the sum of the observed values divided by the number of categories (see note below)

$$\frac{400}{4} = 100$$

Step 2: Calculate O – E

The difference between the observed and expected values is calculated as a measure of the deviation from a predicted result. Since some deviations are negative, they are all squared to give positive values. This step is usually performed as part of a tabulation (right, darker gray column).

Category	O	E	O – E	(O – E)²	$\frac{(O-E)^2}{E}$
Gray, long wing	98	100	–2	4	0.04
Gray, vestigial wing	88	100	–12	144	1.44
Ebony, long wing	102	100	2	4	0.04
Ebony, vestigial wing	112	100	12	144	1.44

Total = 400 χ^2 Σ = 2.96

Step 3: Calculate the value of χ^2

$$\chi^2 = \sum \frac{(O-E)^2}{E}$$

Where: O = the observed result
E = the expected result
Σ = sum of

The calculated χ^2 value is given at the bottom right of the last column in the tabulation.

Step 5a: Using the χ^2 table

On the χ^2 table (part reproduced in Table 1 below) with 3 degrees of freedom, the calculated value for χ^2 of 2.96 corresponds to a probability of between 0.2 and 0.5 (see arrow). *This means that by chance alone a χ^2 value of 2.96 could be expected between 20% and 50% of the time.*

Step 4: Calculating degrees of freedom

The probability that any particular χ^2 value could be exceeded by chance depends on the number of degrees of freedom. This is simply **one less than the total number of categories** (this is the number that could vary independently without affecting the last value). **In this case: 4–1 = 3**.

Step 5b: Using the χ^2 table

The probability of between 0.2 and 0.5 is higher than the 0.05 value which is generally regarded as significant. The null hypothesis cannot be rejected and we have no reason to believe that the observed results differ significantly from the expected (at $P = 0.05$).

Footnote: Many Mendelian crosses involve ratios other than 1:1. For these, calculation of the expected values is not simply a division of the total by the number of categories. Instead, the total must be apportioned according to the ratio. For example, for a total of 400 as above, in a predicted 9:3:3:1 ratio, the total count must be divided by 16 (9+3+3+1) and the expected values will be 225: 75: 75: 25 in each category.

Table 1: Critical values of χ^2 at different levels of probability. By convention, the critical probability for rejecting the null hypothesis (H_0) is 5%. If the test statistic is less than the tabulated critical value for $P = 0.05$ we cannot reject H_0 and the result is not significant. If the test statistic is greater than the tabulated value for $P = 0.05$ we reject H_0 in favor of the alternative hypothesis.

Degrees of freedom	Level of probability (P)									
	0.98	0.95	0.80	0.50	0.20	0.10	0.05	0.02	0.01	0.001
1	0.001	0.004	0.064	0.455	1.64	2.71	3.84	5.41	6.64	10.83
2	0.040	0.103	0.466	1.386	3.22	4.61	5.99	7.82	9.21	13.82
3	0.185	0.352	1.005	2.366	4.64	6.25	7.82	9.84	11.35	16.27
4	0.429	0.711	1.649	3.357	5.99	7.78	9.49	11.67	13.28	18.47
5	0.752	0.145	2.343	4.351	7.29	9.24	11.07	13.39	15.09	20.52

← Do not reject H_0 Reject H_0 →

Chi-Squared Exercise in Genetics

The following problems examine the use of the chi-squared (χ^2) test in genetics. A worked example illustrating the use of the chi-squared test for a genetic cross is provided on the previous page.

1. In a tomato plant experiment, two heterozygous individuals were crossed (the details of the cross are not relevant here). The predicted Mendelian ratios for the offspring of this cross were **9:3:3:1** for each of the **four following phenotypes**: purple stem-jagged leaf edge, purple stem-smooth leaf edge, green stem-jagged leaf edge, green stem-smooth leaf edge.

   The observed results of the cross were not exactly as predicted.
   The numbers of offspring with each phenotype are provided below:

Observed results of the tomato plant cross			
Purple stem-jagged leaf edge	12	Green stem-jagged leaf edge	8
Purple stem-smooth leaf edge	9	Green stem-smooth leaf edge	0

   (a) State your null hypothesis for this investigation (H_0): _____

   (b) State the alternative hypothesis (H_A): _____

2. Use the chi-squared (χ^2) test to determine if the differences observed between the phenotypes are significant. The table of critical values of χ^2 at different P values is provided on the previous pages.

   (a) Enter the observed values (number of individuals) and complete the table to calculate the χ^2 value:

Category	O	E	O – E	(O – E)^2	$\frac{(O-E)^2}{E}$
Purple stem, jagged leaf					
Purple stem, smooth leaf					
Green stem, jagged leaf					
Green stem, smooth leaf					
	Σ				Σ

   (b) Calculate χ^2 value using the equation:

   $$\chi^2 = \sum \frac{(O-E)^2}{E} \qquad \chi^2 = \underline{\qquad}$$

   (c) Calculate the degrees of freedom: _____

   (d) Using the χ^2 table, state the P value corresponding to your calculated χ^2 value:
   _____

   (e) State your decision: reject H_0 / do not reject H_0
   (circle one)

3. Students carried out a pea plant experiment, where two heterozygous individuals were crossed. The predicted Mendelian ratios for the offspring were **9:3:3:1** for each of the **four following phenotypes**: round-yellow seed, round-green seed, wrinkled-yellow seed, wrinkled-green seed.

   The observed results were as follows:

Round-yellow seed	441	Wrinkled-yellow seed	143
Round-green seed	159	Wrinkled-green seed	57

   Use a separate piece of paper to complete the following:
   (a) State the null and alternative hypotheses (H_0 and H_A).
   (b) Calculate the χ^2 value.
   (c) Calculate the degrees of freedom and state the P value corresponding to your calculated χ^2 value.
   (d) State whether or not you reject your null hypothesis: reject H_0 / do not reject H_0 (circle one)

4. Comment on the whether the χ^2 values obtained above are similar. Suggest a reason for any difference:
   _____
   _____

The Structure of a Report

Once you have collected and analysed your data, you can write your report. You may wish to present your findings as a written report, a poster presentation, or an oral presentation. The structure of a scientific report is described below using a poster presentation (which is necessarily very concise) as an example. When writing your report, it is useful to write the methods or the results first, followed by the discussion and conclusion. Although you should do some reading in preparation, the introduction should be one of the last sections that you write. Writing the other sections first gives you a better understanding of your investigation within the context of other work in the same area.

To view this and other examples of posters, see the excellent NC State University web site listed below

1. Title (and author)
Provides a clear and concise description of the project.

2. Introduction
Includes the aim, hypothesis, and background to the study

3. Materials and Methods
A description of the materials and procedures used.

4. Results
An account of results including tables and graphs. This section should not discuss the result, just present them.

5. Discussion
An discussion of the findings in light of the biological concepts involved. It should include comments on any limitations of the study.

6. Conclusion
A clear statement of whether tor not the findings support the hypothesis. In abbreviated poster presentations, these sections may be combined.

7. References & acknowledgements
An organised list of all sources of information. Entries should be consistent within your report. Your teacher will advise you as to the preferred format.

Poster content:

Flounder Exhibit Temperature-Dependent Sex Determination
J. Adam Luckenbach*, John Godwin and Russell Borski
Department of Zoology, Box 7617, North Carolina State University, Raleigh, NC 27695

Introduction
Southern flounder (*Paralichthys lethostigma*) support valuable fisheries and show great promise for aquaculture. Female flounder are known to grow faster and reach larger adult... Therefore, information on sex det... might increase the ratio of female... important for aquaculture.

Objective
This study was conducted to determine whether southern flounder exhibit temperature-dependent sex determination (TSD), and if growth is affected by rearing temperature.

Methods
- Southern flounder broodstock were strip spawned to collect eggs and sperm for *in vitro* fertilization.
- Hatched larvae were weaned from a natural diet (rotifers/*Artemia*) to high protein pelleted feed and fed until satiation at least twice daily.
- Upon reaching a mean total length of 40 mm, the juvenile flounder were stocked at equal densities into one of three temperatures 18, 23, or 28°C for 245 days.
- Gonads were preserved and later sectioned at 2-6 microns.
- Sex-distinguishing markers were used to distinguish males (spermatogenesis) from females (oogenesis).

Histological Analysis
Male Differentiation — Female Differentiation

Results
- Sex was discernible in most fish greater than 120 mm long.
- High (28°C) temperature produced 4% females.
- Low (18°C) temperature produced 22% females.
- Mid-range (23°C) temperature produced 44% females.
- Fish raised at high or low temperatures showed reduced growth compared to those at the mid-range temperature.
- Up to 245 days, no differences in growth existed between sexes.

Conclusions
- These findings indicate that sex determination in southern flounder is temperature-sensitive and temperature has a profound effect on growth.
- A mid-range rearing temperature (23°C) appears to maximize the number of females and promote better growth in young southern flounder.
- Although adult females are known to grow larger than males, no difference in growth between sexes occurred in age-0 (< 1 year) southern flounder.

Acknowledgements
The authors acknowledge the Salstonstall-Kennedy Program of the National Marine Fisheries Service and the University of North Carolina Sea Grant College Program for funding this research. Special thanks to Lea Ware and Beth Shimps for help with the work.

Image courtesy: Adam Luckenbach, NC State University

---

1. Explain the purpose of each of the following sections of a report. The first has one been completed for you:

    (a) Introduction: _Provides the reader with the background to the topic and the rationale for the study_

    (b) Methods: _____

    (c) Results: _____

    (d) Discussion: _____

    (e) References and acknowledgements: _____

2. Posters are a highly visual method of presenting the findings of a study. Describe the positive features of this format:

Writing the Methods

The materials and methods section of your report should be brief but informative. All essential details should be included but those not necessary for the repetition of the study should be omitted. The following diagram illustrates some of the important details that should be included in a methods section. Obviously, a complete list of all possible equipment and procedures is not possible because each experiment or study is different. However, the sort of information that is required for both lab and field based studies is provided.

Field Studies

Study site & organisms
- Site location and features
- Why that site was chosen
- Species involved

Specialized equipment
- pH and oxygen meters
- Thermometers
- Nets and traps

Data collection
- Number and timing of observations/collections
- Time of day or year
- Sample sizes and size of the sampling unit
- Methods of preservation
- Temperature at time of sampling
- Weather conditions on the day(s) of sampling
- Methods of measurement/sampling
- Methods of recording

Laboratory Based Studies

Data collection
- Pre-treatment of material before experiments
- Details of treatments and controls
- Duration and timing of experimental observations
- Temperature
- Sample sizes and details of replication
- Methods of measurement or sampling
- Methods of recording

Experimental organisms
- Species or strain
- Age and sex
- Number of individuals used

Specialized equipment
- pH meters
- Water baths & incubators
- Spectrophotometers
- Centrifuges
- Aquaria & choice chambers
- Microscopes and videos

Special preparations
- Techniques for the preparation of material (staining, grinding)
- Indicators, salt solutions, buffers, special dilutions

General guidelines for writing a methods section

- Choose a suitable level of detail. *Too little detail and the study could not be repeated. Too much detail obscures important features.*
- Do NOT include the details of standard procedures (e.g. how to use a balance) or standard equipment (e.g. beakers and flasks).
- Include details of any statistical analyses and data transformations.
- Outline the reasons why procedures were done in a certain way or in a certain order, if this is not self-evident.
- If your methodology involves complicated preparations (e.g. culture media) then it is acceptable to refer just to the original information source (e.g. lab manual) or include the information as an appendix.

1. The following text is part of the methods section from a report. Using the information above and on the checklist on page 55, describe eight errors (there are ten) in the methods. The errors are concerned with a lack of explanation or detail that would be necessary to repeat the experiment (they are not typographical, nor are they associated with the use of the active voice, which is now considered preferable to the passive):

   "We collected the worms for this study from a pond outside the school. We carried out the experiment at room temperature on April 16, 2004. First we added full strength seawater to each of three 200 cm³ glass jars; these were the controls. We filled another three jars with diluted seawater. We blotted the worms dry and weighed them to the nearest 0.1 g, then we added one worm to each jar. We reweighed the worms (after blotting) at various intervals over the next two hours."

   (a) _____
   (b) _____
   (c) _____
   (d) _____
   (e) _____
   (f) _____
   (g) _____
   (h) _____

Writing Your Results

The results section is arguably the most important part of any research report; it is the place where you can bring together and present your findings. When properly constructed, this section will present your results clearly and in a way that shows you have organized your data and carefully considered the appropriate analysis. A portion of the results section from a scientific paper on the habitat preference of New Zealand black mudfish is presented below (Hicks, B. and Barrier, R. (1996), NZJMFR. 30, 135-151). It highlights some important features of the results section and shows you how you can present information concisely, even if your results are relatively lengthy. Use it as a guide for content when you write up this section.

Results

A total of 222 black mudfish were caught in the 400 traps set b[...] Mean total length (TL) was 67 mm (range 26-139 mm, $n = 214$)[...] had black mudfish. Mudfi[...] ly amo[...] independence, $P < 0.001$: [...] at 8 out [...] at 20 out of 30 wetland sit[...] at only [...] none of the 6 lake margin or 4 pond, dam, and lagoon sites. Categorical variables that distinguished [...] χ^2 tests of independence, $P < 0.05$: Table 4) were: [...] ate disturbance scale rating; presence of emergent [...] sed or peat bog substrate types; absence of fish [...] *orphus cotidianus*) and inanga (*Galaxias maculatus*);

> Graphs (figures) illustrate trends in the data. Be sure to choose the correct type of graph and allocate enough space to it in the report.

> Keep your statement of important findings brief.

> Label figures and tables clearly and in sequence so that they can be referred to easily in the text.

> Scientific names are included if they are known.

Table 4: χ^2 tests of association between presence or absence of black mudfish and categorical habitat variables at sites in the Waikato region.

Habitat variable	χ^2 statistic	df	Probability	
Absence of water in summer	31.84	1	<0.001	
Disturbance scale rating	23.92	4	<0.001	
Emergent vegetation	17.56	1	<0.001	
Overhanging vegetation	11.82	1	<0.001	Significant
Surface substrate type	16.51	2	<0.001	
Absence of bullies and inanga	6.17	1	0.013	
Tree roots	4.00	1	0.040	
Underlying soil type	8.05	4	0.090	Not significant

> Tables summarize raw data, any transformations, and the results of statistical tests.

> Distinction is made between those statistical values that are significant and those that are not (at $P < 0.05$).

Means of t[...] depth, wint[...] water depth[...] es with and [...] 5). Mean summer water depth was only 2.1 cm at sites with mudfish, compared to 22.6 cm at sites without. Winter and maximum water depths were also less at sites with mudfish than at sites without mudfish. M[...] elometric turbidity units (NTU) at sites with mudfish, but 21.3 NTU at site[...] ity, total dissolved solids, dissolved oxygen at the water surface, humic cond[...] ocity were similar at all sites (t-test, $P > 0.153$: Table 5). Catch rates at site[...] mudfish per trap per night (geometric mean 0.70: Table 5).

> Only include results; this is not the place to discuss them.

> Any abbreviations are noted the first time they occur.

> Tables and figures are referred to in brackets.

Table 5: Characteristics [...]

| Variable | Sites with mudfish | | Sites without mudfish | | Proba[...] |
	Mean ± CI	SD	Mean ± CI	SD	
Mudfish catch rate /fish per trap per night	0.70 ± 1.31	0.84	0.00		
Summer water depth /cm	2.1 ± 1.9	2.2	22.6 ± 7.8	24.7	<0.001
Winter water depth /cm	28.9 ± 4.3	5.8	40.2 ± 7.7	24.4	0.012
Turbidity /NTU	11.5 ± 2.5	13.3	21.3 ± 7.1	22.6	0.012

Catch rates for classes within variables, and the habitat preference [...] that mudfish were virtually absent from water of > 30 cm depth in s[...] mudfish preferred water depths between 15 and 50 cm. Disturban[...] preferred, as were turbidities of < 15 NTU. Preference for the DSR of 1 was assumed to be the same as for the DSR of 2, as the small number of sites with DSR of 1 and 2 ($n = 5$ in each case) made their separate preferences unreliable.

> Any extra information for a figure can be shown by an asterisk and included elsewhere.

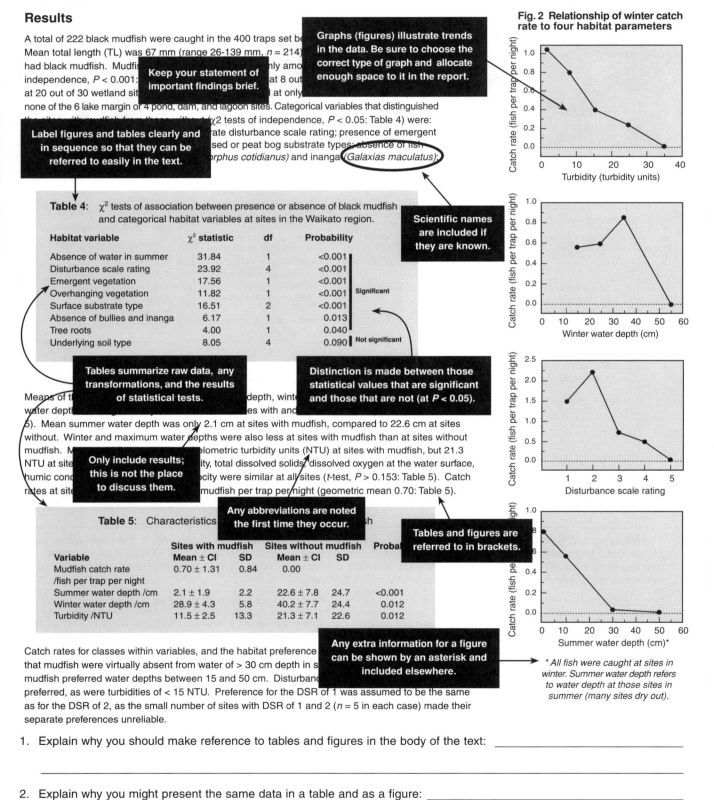

Fig. 2 Relationship of winter catch rate to four habitat parameters

* All fish were caught at sites in winter. Summer water depth refers to water depth at those sites in summer (many sites dry out).

1. Explain why you should make reference to tables and figures in the body of the text: _____

2. Explain why you might present the same data in a table and as a figure: _____

> # Writing Your Discussion

In the discussion section of your report, you must interpret your results in the context of the specific questions you set out to answer in the investigation. You should also place your findings in the context of any broader relevant issues. If your results coincide exactly with what you expected, then your discussion will be relatively brief. However, be prepared to discuss any unexpected or conflicting results and critically evaluate any problems with your study design. The Discussion section may (and should) refer to the findings in the Results section, but it is not the place to introduce new results. Try to work towards a point in your discussion where the reader is lead naturally to the conclusion. The conclusion may be presented within the discussion or it may be included separately after the discussion as a separate section.

Discussion:

Black mudfish habitat in the Waikato region can be adequately characterised for analyses by four variables that are easy to measure: summer water depth, winter water depth, DSR (modification indicated by vegetation), and turbidity. Catch rates of black mudfish can be extremely high. In our study, catch rates ranged from 0.2 to 8.4 mudfish per trap per night (mean 0.70) between May and October 1992, and were similar to those of Dean (1995) in September 1993 and October 1994 in the Whangamarino Wetland complex (0.0-2.0 mudfish per trap per night). The highest mean catch rate in our study, 8.4 mudfish per trap per night, was at Site 24 (Table 1, Figure 1). The second highest (6.4 mudfish per trap per night) was at Site 32, in a drain about 4 km east of Hamilton. Black mudfish in the Waikato region were most commonly found at sites in wetlands with absence of water in summer, moderate depth of water in winter, limited modification of the vegetation (low DSR), and low turbidity (Fig. 2). There are similarities between the habitat requirements of black mudfish and those of brown mudfish and the common river galaxias *(Galaxias vulgaris)*. Brown mudfish inhabited shallow water, sometimes at the edges of deeper water bodies, but were usually absent from water deeper than about 30-50 cm (Eldon 1978). The common river galaxias also has a preference for shallow water, occupying river margins < 20 cm deep (Jowett and Richardson 1995).

Sites where black mudfish were found were not just shallow or dry in summer, but showed seasonal variation in water depth. A weakness of this study is the fact that sites were trapped only once. Traps were spread relatively widely at each site to maximise the chance of catching any fish present. Cover was important for black mudfish, in the form of emergent or overhanging vegetation, or tree roots. The significance of cover in determining the presence of black mudfish was predictable, considering the shallow nature of their habitats. Mudfish, though nocturnal in their activity, require cover during the to protect them from avian predators, such as bitterns *(Botaurus poiciloptilus)* and kingfishers *(Halcyon sancta vagans)*. Predation of black mudfish by a swamp bittern has been recorded (Ogle 1981). Cover is also important for brown mudfish (Eldon 1978). Black mudfish were found at sites with the predatory mosquitofish and juvenile eels, and the seasonal drying of their habitats may be a key to the successful coexistence of mudfish with their predators. Mosquitofish are known predators of mudfish fry (Barrier & Hicks 1994), and eels would presumably also prey on black mudfish, as they do on brown mudfish (Eldon 1979b). If, however, black mudfish are relatively uncompetitive and vulnerable to predation, questions remain as to how they manage to coexist with juvenile eels and mosquitofish. The habitat variables measured in this study can be used to classify the suitability of sites for black mudfish in future. The adaptability of black mudfish allows them to survive in some altered habitats, such as farm or roadside drains. From this study, we can conclude that the continued existence of suitable habitats appears to be more important to black mudfish than the presence of predators and competitors. This study has also improved methods of identifying suitable mudfish habitats in the Waikato region.

Annotations:
- Support your statements with reference to Tables and Figures from the Results section.
- The discussion describes the relevance of the results of the investigation.
- State any limitations of your approach in carrying out the investigation and what further studies might be appropriate.
- Reference is made to the work of others.
- Further research is suggested
- A clear conclusion is made towards the end of the discussion.

1. Explain why it is important to discuss any weaknesses in your study design: _____

2. Explain why you should **critically evaluate** your results in the discussion: _____

3. Describe the purpose of the conclusion: _____

Report Checklist

A report of your findings at the completion of your investigation may take one of the following forms: a written document, seminar, poster, web page, or multimedia presentation. The following checklist identifies some important points to consider when writing each section of your report. Review the list before you write your report and then, once you have successfully completed each section of your write-up, use the check boxes to tick off the points.

Title:
- ☐ (a) Gives a clear indication of what the study is about.
- ☐ (b) Includes the species name and a common name of all organisms used.

Introduction:
- ☐ (a) Includes a clear aim.
- ☐ (b) Includes a well written hypothesis.
- ☐ (c) Includes a synopsis of the current state of knowledge about the topic.

Materials and methods:
- ☐ (a) Written clearly. Numbered points are appropriate at this level.
- ☐ (b) Describes the final methods that were used.
- ☐ (c) Includes details of the how data for the dependent variable were collected.
- ☐ (d) Includes details of how all other variables were manipulated, controlled, measured, or monitored.
- ☐ (e) If appropriate, it includes an explanatory diagram of the design of the experimental set-up.
- ☐ (f) Written in the past tense, and in the active voice (We investigated …) rather than the passive voice (An investigation was done …).

Results:
- ☐ (a) Includes the raw data (e.g. in a table).
- ☐ (b) Where necessary, the raw data have been averaged or transformed.
- ☐ (c) Includes graphs (where appropriate).
- ☐ (d) Each figure (table, graph, drawing, or photo) has a title and is numbered in a way that makes it possible to refer to it in the text (Fig. 1 etc.).
- ☐ (e) Written in the past tense and, where appropriate, in the active voice.

Discussion:
- ☐ (a) Includes an analysis of the data in which the findings, including trends and patterns, are discussed in relation to the biological concepts involved.
- ☐ (b) Includes an evaluation of sources of error, assumptions, and possible improvements to design.

Conclusion:
- ☐ (a) Written as a clear statement, which relates directly to the hypothesis.

Bibliography or References:
- ☐ (a) Lists all sources of information and assistance.
- ☐ (b) Does not include references that were not used.

Related activities: The Structure of a Report

Citing and Listing References

Proper referencing of sources of information is an important aspect of report writing. It shows that you have explored the topic and recognise and respect the work of others. There are two aspects to consider: **citing sources** within the text (making reference to other work to support a statement or compare results) and **compiling a reference list** at the end of the report. A **bibliography** lists all sources of information, but these may not necessarily appear as citations in the report. In contrast, a reference list should contain only those texts cited in the report.

Citations in the main body of the report should include only the authors' surnames, publication date, and page numbers (or internet site) and the citation should be relevant to the statement it claims to support. Accepted methods for referencing vary, but your reference list should provide all the information necessary to locate the source material, it should be consistently presented, and it should contain only the references that you have *yourself* read (not those cited by others). A suggested format using the **APA** referencing system is described below.

Preparing a Reference List

When teachers ask students to write in "APA style", they are referring to the editorial style established by the **American Psychological Association** (APA). These guidelines for citing **electronic (online) resources** differ only slightly from the **print sources**.

For the Internet

Where you use information from the internet, you must provide the following:
- The website address (URL), the person or organization who is in charge of the web site and the date you accessed the web page.

This is written in the form: URL (person or organization's name, day, month, and year retrieved)
This goes together as follows:
> http://www.scientificamerican.com (Scientific American, 17.12.03)

For Periodicals (or Journals)

This is written in the form: author(s), date of publication, article title, periodical title, and publication information.
Example: Author's family name, A. A. (author's initials only), Author, B. B., & Author, C. C. (xxxx = year of publication in brackets). Title of article. Title of Periodical, volume number, page numbers (Note, only use "pp." before the page numbers in newspapers and magazines).
This goes together as follows:
> Bamshad M. J., & Olson S. E. (2003). Does Race Exist? Scientific American, 289(6), 50-57.

For Online Periodicals based on a Print Source

At present, the majority of periodicals retrieved from online publications are exact duplicates of those in their print versions and although they are unlikely to have additional analyses and data attached to them, this is likely to change in the future.

- If the article that is to be referenced has been viewed only in electronic form and not in print form, then you must add in brackets, "Electronic version", after the title.
This goes together as follows:
> Bamshad M. J., & Olson S. E. (2003). Does Race Exist? (Electronic version). Scientific American, 289(6), 50-57.

- If you have reason to believe the article has changed in its electronic form, then you will need to add the date you retrieved the document and the URL.
This goes together as follows:
> Bamshad M. J., & Olson S. E. (2003). Does Race Exist? (Electronic version). Scientific American, 289(6), 50-57. Retrieved December 17, 2003, from http://www.scientificamerican.com

For Books

This is written in the form: author(s), date of publication, title, and publication information.
Example: Author, A. A., Author, B. B., & Author, C. C. (xxxx). Title (any additional information to enable identification is given in brackets). City of publication: publishers name.
This goes together as follows:
> Martin, R.A. (2004). Missing Links Evolutionary Concepts & Transitions Through Time. Sudbury, MA: Jones and Bartlett

For Citation in the Text of References

This is written in the form: authors' surname(s), date of publication, page number(s) (abbreviated p.), chapter (abbreviated chap.), figure, table, equation, or internet site, in brackets at the appropriate point in text.
This goes together as follows:
> (Bamshad & Olson, 2003, p. 51) or (Bamshad & Olson, 2003, http://www.scientificamerican.com)

This can also be done in the form of footnotes. This involves the use of a superscripted number in the text next to your quoted material and the relevant information listed at the bottom of the page.
This goes together as follows:
> .... Bamshad & Olson reported that ....[1]

---
[1] Bamshad & Olson, 2003, p. 51

Example of a Reference List

Lab notes can be listed according to title if the author is unknown.
→ Advanced biology laboratory manual (2000). Cell membranes. pp. 16-18. Sunhigh College.

References are listed alphabetically according to the author's surname.

Cooper, G.M. (1997). *The cell: A molecular approach* (2nd ed.). Washington D.C.: ASM Press
   — Book title in italics (or underlined)
   — Place of publication: Publisher

Davis, P. (1996). Cellular factories. *New Scientist* 2057: Inside science supplement.
   — Publication date
   — Journal title in italics
   — A supplement may not need page references

If a single author appears more than once, then list the publications from oldest to most recent.

Indge, B. (2001). Diarrhea, digestion and dehydration. *Biological Sciences Review,* 14(1), 7-9.

Indge, B. (2002). Experiments. *Biological Sciences Review,* 14(3), 11-13.
   — Article title follows date

Kingsland, J. (2000). Border control. *New Scientist* 2247: Inside science supplement.

Laver, H. (1995). Osmosis and water retention in plants. *Biological Sciences Review,* 7(3), 14-18.
   — Volume (Issue number), Pages

Spell out only the last name of authors. Use initials for first and middle names.

Steward, M. (1996). Water channels in the cell membrane. *Biological Sciences Review,* 9(2), 18-22.

Internet sites change often so the date accessed is included. The person or organization in charge of the site is also included.
→ http://www.cbc.umn.edu/~mwd/cell_intro.html (Dalton, M. "Introduction to cell biology" 12.02.03)

---

1. Distinguish between a **reference list** and a **bibliography**:

2. Explain why internet articles based on a print source are likely to have additional analyses and data attached in the future, and why this point should be noted in a reference list:

3. Following are the details of references and source material used by a student in preparing a report on enzymes and their uses in biotechnology. He provided his reference list in prose. From it, compile a correctly formatted reference list:

Pages 18-23 in the sixth edition of the textbook "Biology" by Neil Campbell. Published by Benjamin/Cummings in California (2002). New Scientist article by Peter Moore called "Fuelled for life" (January 1996, volume 2012, supplement). "Food biotechnology" published in the journal Biological Sciences Review, page 25, volume 8 (number 3) 1996, by Liam and Katherine O'Hare. An article called "Living factories" by Philip Ball in New Scientist, volume 2015 1996, pages 28-31. Pages 75-85 in the book "The cell: a molecular approach" by Geoffrey Cooper, published in 1997 by ASM Press, Washington D.C. An article called "Development of a procedure for purification of a recombinant therapeutic protein" in the journal "Australasian Biotechnology", by I Roberts and S. Taylor, pages 93-99 in volume 6, number 2, 1996.

REFERENCE LIST

Field Studies

Developing scientific investigative skills and attitudes in the field

Designing a field study. Sampling populations. Measuring abiotic factors. Gathering and systematically recording field data.

Learning Objectives

☐ 1. Compile your own glossary from the **KEY WORDS** displayed in **bold type** in the learning objectives below.

Designing a Field Study *(pages 65-66)*

☐ 2. Use the general guidelines and procedures provided in the topic *Biological Investigations* to plan a field study to investigate an aspect of a population or community.

Sampling Populations *(pages 64, 67-82)*

☐ 3. Explain what is meant by a **sample** as it relates to studies of populations or communities. Explain why we sample populations and describe the advantages and drawbacks involved. Recognize that the design of your field study should enable you to obtain reliable information about your population or community.

☐ 4. Develop an understanding of the terminology involved in field studies. You should distinguish between:
(a) A data value for a particular variable (e.g. length or percentage plant cover).
(b) The individual sampling unit (e.g. an individual limpet or a quadrat of a particular size).
(c) The sample size (*n*), i.e. the number of sampling units (e.g. 50 limpets or the number of quadrats).
(d) The population (e.g. the limpets on a rocky shore, or the total number of plants in a field).

☐ 5. Appreciate the range of techniques available for collecting data from populations and communities: direct counts, **frame** and/or **point quadrats**, **belt** and/or **line transects**, **mark and recapture**, **netting** and **trapping**, **indirect methods** (e.g. recording calls or scat), and **radio-tracking**. Identify the advantages and limitations of each method. Use this information to select an appropriate sampling method for the population or community you wish to study.

☐ 6. Collect and **systematically record** field data from a named population or community. Consider the following:
   ☐ The size of the sampling unit (e.g. quadrat size). Discuss how you would decide on a suitable **sample size** and (if appropriate) **size of sampling unit**. Appreciate how and why these might affect the reliability of the data you collect.
   ☐ The number of sampling units in each sample (e.g. the number of quadrats you will use).
   ☐ The location of sampling units in the sampling area (e.g. random placement or systematic sampling with a random start point). Describe how you would ensure **random sampling**, and why this is important.

☐ 7. Measure aspects of the physical environment that might be relevant to your aim or important in explaining any population or community pattern or interrelationship.

☐ 8. Recognize the compromise existing between sampling accuracy and sampling effort. Check that your field study relates to the aims of your investigation, that it is practical and achievable, and that it provides representative, unbiased information about the population or community under study. For this, you will need to collect **sufficient data**.

Recording and Presenting your Findings

☐ 9. Using the guidelines provided in the topic *Biological Investigations*, **systematically process** the data collected from your field study in an appropriate way.

☐ 10. With reference to environmental factors, discuss any pattern or interrelationship indicated by your results. Consider the original aim of your investigation in the light of your results and your discussion of them.

See page 7 for additional details of these texts:
■ Adds, J. *et al.*, 1999. **Tools, Techniques and Assessment in Biology** (NelsonThornes).
■ Indge, B., 2003. **Data and Data Handling for AS and A Level Biology** (Hodder Arnold H&S).
■ Jones, A., *et al.*, 2007. **Practical Skills in Biology** (Addison-Wesley), as required.

Presentation MEDIA to support this topic:
ECOLOGY:
• Practical Ecology
• Communities

See page 7 for details of publishers of periodicals:
■ **Fieldwork - Sampling Animals** Biol. Sci. Rev., 10(4) March 1998, pp. 23-25. *Excellent article covering the appropriate methodology for collecting different types of animals in the field. Includes an excellent synopsis of the mark and recapture technique.*

■ **Fieldwork Sampling - Plants** Biol. Sci. Rev., 10(5) May 1998, pp. 6-8. *Methodology for sampling plant communities. Includes thorough coverage of quadrat use.*

■ **Bird Ringing** Biol. Sci. Rev., 14(3) Feb. 2002, pp. 14-19. *Techniques used in investigating populations of highly mobile organisms: mark and recapture methods, ringing techniques, and application of diversity indices.*

■ **Bowels of the Beasts** New Scientist, 22 August 1998, pp. 36-39. *Analyses of the faeces of animals can reveal much about the make-up, size, and genetic diversity of a population.*

See pages 4-5 for details of how to access **Bio Links** from our web site: **www.thebiozone.com** From Bio Links, access sites under the topics:

ECOLOGY: > **Environmental Monitoring**:
• Amphibian monitoring program • Remote sensing and monitoring > **Populations and Communities:** • Communities • Quantitative population ecology • Sirtracking for wildlife research ... *and others*

Sampling Populations

Information about the populations of rare organisms in isolated populations may, in some instances, be collected by direct measure (direct counts and measurements of all the individuals in the population). However, in most cases, populations are too large to be examined directly and they must be sampled in a way that still provides information about them. Most practical exercises in population ecology involve the collection of living organisms, with a view to identifying the species and quantifying their abundance and other population features of interest. Sampling techniques must be appropriate to the community being studied and the information you wish to obtain. Some of the common strategies used in ecological sampling, and the situations for which they are best suited, are outlined in the table below. It provides an overview of points to consider when choosing a sampling regime. One must always consider the time and equipment available, the organisms involved, and the impact of the sampling method on the environment. For example, if the organisms involved are very mobile, sampling frames are not appropriate. If it is important not to disturb the organisms, observation alone must be used to gain information.

Method	Equipment and procedure	Information provided and considerations for use
Point sampling Random / Systematic (grid)	Individual points are chosen on a map (using a grid reference or random numbers applied to a map grid) and the organisms are sampled at those points. Mobile organisms may be sampled using traps, nets etc.	**Useful for**: Determining species abundance and community composition. If samples are large enough, population characteristics (e.g. age structure, reproductive parameters) can be determined. **Considerations**: Time efficient. Suitable for most organisms. Depending on methods used, disturbance to the environment can be minimized. Species occurring in low abundance may be missed.
Transect sampling 0.5 m, Environmental gradient	Lines are drawn across a map and organisms occurring along the line are sampled. **Line transects**: Tape or rope marks the line. The species occurring on the line are recorded (all along the line or, more usually, at regular intervals). Lines can be chosen randomly (left) or may follow an environmental gradient. **Belt transects**: A measured strip is located across the study area to highlight any transitions. Quadrats are used to sample the plants and animals at regular intervals along the belt. Plants and immobile animals are easily recorded. Mobile or cryptic animals need to be trapped or recorded using appropriate methods.	**Useful for**: Well suited to determining changes in community composition along an environmental gradient. When placed randomly, they provide a quick measure of species occurrence. **Considerations for line transects**: Time efficient. Most suitable for plants and immobile or easily caught animals. Disturbance to the environment can be minimized. Species occurring in low abundance may be missed. **Considerations for belt transects**: Time consuming to do well. Most suitable for plants and immobile or easily caught animals. Good chance of recording most or all species. Efforts should be made to minimize disturbance to the environment.
Quadrat sampling	Sampling units or quadrats are placed randomly or in a grid pattern on the sample area. The occurrence of organisms in these squares is noted. Plants and slow moving animals are easily recorded. Rapidly moving or cryptic animals need to be trapped or recorded using appropriate methods.	**Useful for**: Well suited to determining community composition and features of population abundance: species density, frequency of occurrence, percentage cover, and biomass (if harvested). **Considerations**: Time consuming to do well. Most suitable for plants and immobile or easily caught animals. Quadrat size must be appropriate for the organisms being sampled and the information required. Some disturbance if organisms are removed.
Mark and recapture (capture-recapture) First sample: marked / Second sample: proportion recaptured	Animals are captured, marked, and then released. After a suitable time period, the population is resampled. The number of marked animals recaptured in a second sample is recorded as a proportion of the total.	**Useful for**: Determining total population density for highly mobile species in a certain area (e.g. butterflies). Movements of individuals in the population can be tracked (especially when used in conjunction with electronic tracking devices). **Considerations**: Time consuming to do well. Not suitable for immobile species. Population should have a finite boundary. Period between samplings must allow for redistribution of marked animals in the population. Marking should present little disturbance and should not affect behavior.

1. Briefly explain why we **sample** populations: _____

2. Describe a sampling technique that would be appropriate for determining each of the following:

   (a) The percentage cover of a plant species in pasture: _____

   (b) The density and age structure of a plankton population: _____

   (c) Change in community composition from low to high altitude on a mountain: _____

Related activities: Quadrat Sampling, Transect Sampling, Mark and Recapture Sampling

Designing Your Field Study

The figure below provides an example and some ideas for designing a field study. It provides a framework which can be modified for most simple comparative field investigations. For reasons of space, the full methodology is not included.

Pill millipede *Glomeris marginata*
Oak woodland | Coniferous woodland

Observation

A student read that a particular species of pill millipede (left) is extremely abundant in forest leaf litter, but a search in the litter of a conifer-dominated woodland near his home revealed only very low numbers of this millipede species.

Hypothesis

This millipede species is adapted to a niche in the leaf litter of oak woodlands and is abundant there. However, it is rare in the litter of coniferous woodland. The **null hypothesis** is that there is no difference between the abundance of this millipede species in oak and coniferous woodland litter.

Oak or coniferous woodland

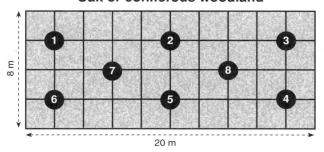

1 Sampling sites numbered 1-8 at evenly spaced intervals on a 2 x 2 m grid within an area of 20 m x 8 m.

Sampling Program

A sampling program was designed to test the **prediction** that the millipedes would be more abundant in the leaf litter of oak woodlands than in coniferous woodlands.

Equipment and Procedure

Sites: For each of the two woodland types, an area 20 x 8 m was chosen and marked out in 2 x 2 m grids. Eight sampling sites were selected, evenly spaced along the grid as shown.

- The general area for the study chosen was selected on the basis of the large amounts of leaf litter present.
- Eight sites were chosen as the largest number feasible to collect and analyze in the time available.
- The two woodlands were sampled on sequential days.

Capture of millipedes: At each site, a 0.4 x 0.4 m quadrat was placed on the forest floor and the leaf litter within the quadrat was collected. Millipedes and other leaf litter invertebrates were captured using a simple gauze lined funnel containing the leaf litter from within the quadrat. A lamp was positioned over each funnel for two hours and the invertebrates in the litter moved down and were trapped in the collecting jar.

- After two hours each jar was labeled with the site number and returned to the lab for analysis.
- The litter in each funnel was bagged, labeled with the site number and returned to the lab for weighing.
- The number of millipedes at each site was recorded.
- The numbers of other invertebrates (classified into major taxa) were also noted for reference.

Sampling equipment: leaf litter light trap

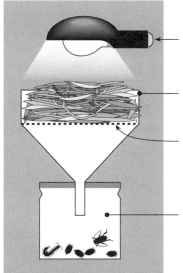

Light from a battery operated lamp drives the invertebrates down through the leaf litter.

Large (diameter 300 mm) funnel containing leaf litter resting on a gauze platform.

Gauze allows invertebrates of a certain size to move down the funnel.

Collecting jar placed in the litter on the forest floor traps the invertebrates that fall through the gauze and prevents their escape.

Assumptions

- The areas chosen in each woodland were representative of the woodland types in terms of millipede abundance.
- Eight sites were sufficient to adequately sample the millipede populations in each forest.
- A quadrat size of 0.4 x 0.4 m contained enough leaf litter to adequately sample the millipedes at each site.
- The millipedes were not preyed on by any of the other invertebrates captured in the collecting jar.
- All the invertebrates within the quadrat were captured.
- Millipedes moving away from the light are effectively captured by the funnel apparatus and cannot escape.
- Two hours was long enough for the millipedes to move down through the litter and fall into the trap.

Note that these last two assumptions could be tested by examining the bagged leaf litter for millipedes after returning to the lab.

Notes on collection and analysis

- Mean millipede abundance was calculated from the counts from the eight sites. The difference in abundance at the sites was tested using a Student's *t* test.
- After counting and analysis of the samples, all the collected invertebrates were returned to the sites.

Sample Size

When designing a field study, the size of your sampling unit (in this case quadrat size) and the sample size (the number of samples you will take) should be major considerations. There are various ways to determine the best quadrat size. Usually, these involve increasing the quadrat size until you stop finding new species. For simple field studies, the number of samples you take (the sample size or n value) will be determined largely by the resources and time available to collect and analyze your data. It is usually best to take as many samples as you can, as this helps to account for any natural variability present and will give you greater confidence in your data. However, at some point (dependent on the data), more samples will not yield much more information.

1. Explain the importance of each of the following in field studies:

    (a) Appropriate quadrat size (or any equivalent sampling unit): _____

    (b) Recognizing any assumptions that you are making: _____

    (c) Appropriate consideration of the environment: _____

    (d) Return of organisms to the same place after removal: _____

    (e) Appropriate size of total sampling area within which the sites are located: _____

2. Describe how you could test whether any given quadrat size was adequate to effectively sample the organism involved: _____

YOUR CHECKLIST FOR FIELD STUDY DESIGN

The following provides a checklist for a field study. Check off the points when you are confident that you have satisfied the requirements in each case:

1. **Preliminary:**

    ☐ (a) Makes a hypothesis based on observation(s).

    ☐ (b) The hypothesis (and its predictions) are testable using the resources you have available (the study is feasible).

    ☐ (c) The organism you have chosen is suitable for the study and you have considered the ethics involved.

2. **Assumptions and site selection:**

    ☐ (a) You are aware of any assumptions that you are making in your study.

    ☐ (b) You have identified aspects of your field design that could present problems (such as time of year, biological rhythms of your test organism, difficulty in identifying suitable habitats etc.).

    ☐ (c) The study sites you have selected have the features necessary in order for you to answer the questions you have asked in your hypothesis.

3. **Data collection:**

    ☐ (a) You are happy with the way in which you are going to take your measurements or samples.

    ☐ (b) You have considered the size of your sampling unit and the number of samples you are going to take (and tested for these if necessary).

    ☐ (c) You have given consideration to how you will analyze the data you collect and made sure that your study design allows you to answer the questions you wish to answer.

Monitoring Physical Factors

Most ecological studies require us to measure the physical factors (parameters) in the environment that may influence the abundance and distribution of organisms. In recent years there have been substantial advances in the development of portable, light-weight meters and dataloggers. These enable easy collection and storage of data in the field.

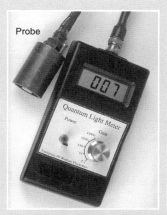

Quantum light meter: Measures light intensity levels. It is not capable of measuring light quality (wavelength).

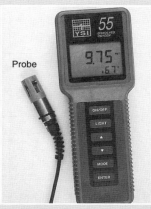

Dissolved oxygen meter: Measures the amount of oxygen dissolved in water (expressed as mgl^{-1}).

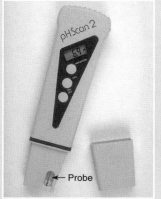

pH meter: Measures the acidity of water or soil, if it is first dissolved in pure water (pH scale 0 to 14).

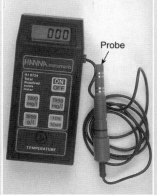

Total dissolved solids (TDS) meter: Measures content of dissolved solids (as ions) in water in mgl^{-1}.

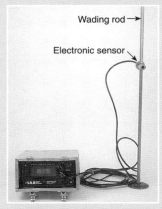

Current meter: The electronic sensor is positioned at set depths in a stream or river on the calibrated wading rod as current readings are taken.

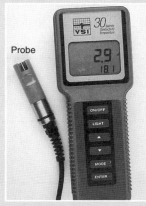

Multipurpose meter: This is a multi-functional meter, which can measure salinity, conductivity and temperature simply by pushing the MODE button.

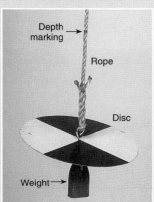

Secchi disc: This simple device is used to provide a crude measure of water clarity (the maximum depth at which the disc can just be seen).

Collecting a water sample: A Nansen bottle is used to collect water samples from a lake for lab analysis, testing for nutrients, oxygen and pH.

Dataloggers and Environmental Sensors

Dataloggers are electronic instruments that record measurements over time. They are equipped with a microprocessor, data storage facility, and sensor. Different sensors are employed to measure a range of variables in water (photos A and B) or air (photos C and D), as well as make physiological measurements. The datalogger is connected to a computer, and software is used to set the limits of operation (e.g. the sampling interval) and initiate the logger. The logger is then disconnected and used remotely to record and store data. When reconnected to the computer, the data are downloaded, viewed, and plotted. Dataloggers, such as those pictured here from PASCO, are being increasingly used in professional and school research. They make data collection quick and accurate, and they enable prompt data analysis.

Dataloggers are now widely used to monitor conditions in aquatic environments. Different variables such as pH, temperature, conductivity, and dissolved oxygen can be measured by changing the sensor attached to the logger.

Dataloggers fitted with sensors are portable and easy to use in a wide range of terrestrial environments. They are used to measure variables such as air temperature and pressure, relative humidity, light, and carbon dioxide gas.

1. The physical factors of an exposed rocky shore and a sheltered estuarine mudflat differ markedly. For each of the factors listed in the table below, briefly describe how they may differ (if at all):

Environmental parameter	Exposed rocky coastline	Estuarine mudflat
Severity of wave action		
Light intensity and quality		
Salinity/ conductivity		
Temperature change (diurnal)		
Substrate/ sediment type		
Oxygen concentration		
Exposure time to air (tide out)		

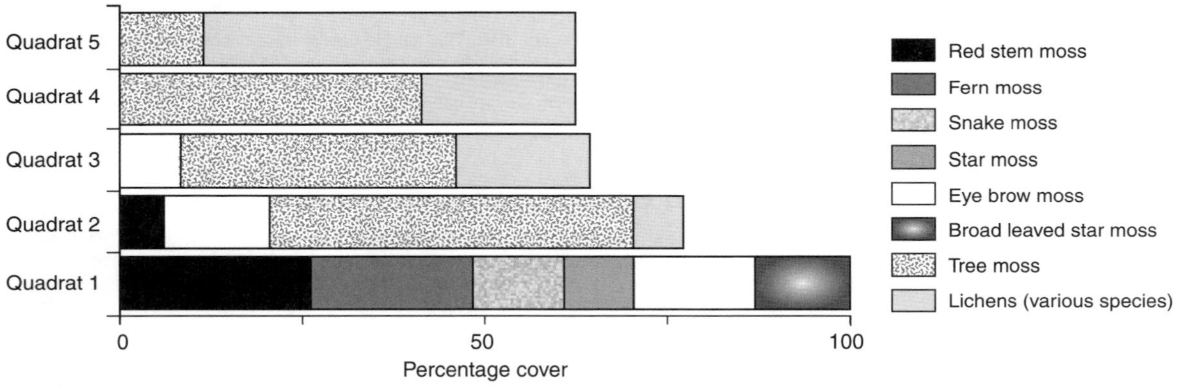

QUADRAT	1	2	3	4	5
Height (m)	0.4	0.8	1.2	1.6	2.0
Light (arbitrary units)	40	56	68	72	72
Humidity (percent)	99	88	80	76	78
Temperature (°C)	12.1	12.2	13	14.3	14.2

2. The figure (above) shows the changes in vegetation cover along a 2 m vertical transect up the trunk of an oak tree (*Quercus*). Changes in the physical factors light, humidity, and temperature along the same transect were also recorded. From what you know about the ecology of mosses and lichens, account for the observed vegetation distribution:

_____

_____

_____

Quadrat Sampling

Quadrat sampling is a method by which organisms in a certain proportion (sample) of the habitat are counted directly. As with all sampling methods, it is used to estimate population parameters when the organisms present are too numerous to count in total. It can be used to estimate population **abundance** (number), **density, frequency of occurrence**, and **distribution**. Quadrats may be used without a transect when studying a relatively uniform habitat. In this case, the quadrat positions are chosen randomly using a random number table.

The general procedure is to count all the individuals (or estimate their percentage cover) in a number of quadrats of known size and to use this information to work out the abundance or percentage cover value for the whole area. The number of quadrats used and their size should be appropriate to the type of organism involved (e.g. grass vs tree).

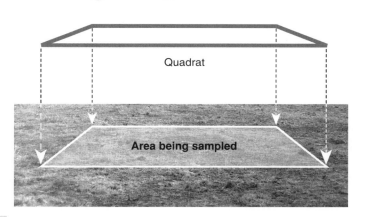

$$\text{Estimated average density} = \frac{\text{Total number of individuals counted}}{\text{Number of quadrats} \times \text{area of each quadrat}}$$

Guidelines for Quadrat Use:

1. The **area of each quadrat** must be known exactly and ideally quadrats should be the same shape. The quadrat does not have to be square (it may be rectangular, hexagonal etc.).
2. **Enough quadrat samples** must be taken to provide results that are representative of the total population.
3. The **population of each quadrat** must be known exactly. Species must be distinguishable from each other, even if they have to be identified at a later date. It has to be decided beforehand what the count procedure will be and how organisms over the quadrat boundary will be counted.
4. The size of the quadrat should be appropriate to the organisms and habitat, e.g. a large size quadrat for trees.
5. The quadrats must be **representative of the whole area**. This is usually achieved by **random sampling** (right).

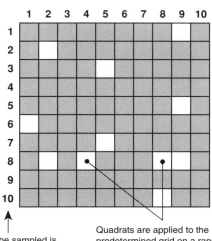

The area to be sampled is divided up into a grid pattern with indexed coordinates

Quadrats are applied to the predetermined grid on a random basis. This can be achieved by using a random number table.

Sampling a centipede population

A researcher by the name of Lloyd (1967) sampled centipedes in Wytham Woods, near Oxford in England. A total of 37 hexagon–shaped quadrats were used, each with a diameter of 30 cm (see diagram on right). These were arranged in a pattern so that they were all touching each other. Use the data in the diagram to answer the following questions.

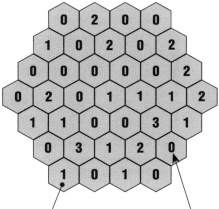

Each quadrat was a hexagon with a diameter of 30 cm and an area of 0.08 square meters.

The number in each hexagon indicates how many centipedes were caught in that quadrat.

1. Determine the average number of centipedes captured per quadrat:

   _____

2. Calculate the estimated average density of centipedes per square metre (remember that each quadrat is 0.08 square metres in area):

   _____

3. Looking at the data for individual quadrats, describe in general terms the distribution of the centipedes in the sample area:

   _____

   _____

4. Describe one factor that might account for the distribution pattern:

   _____

   _____

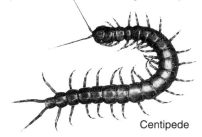

Centipede

Related activities: Quadrat-Based Estimates, Sampling a Leaf Litter Population

Quadrat-Based Estimates

The simplest description of a plant community in a habitat is a list of the species that are present. This qualitative assessment of the community has the limitation of not providing any information about the **relative abundance** of the species present. Quick estimates can be made using **abundance scales**, such as the ACFOR scale described below. Estimates of percentage cover provide similar information. These methods require the use of **quadrats**. Quadrats are used extensively in plant ecology. This activity outlines some of the common considerations when using quadrats to sample plant communities.

What Size Quadrat?

Quadrats are usually square, and cover 0.25 m^2 (0.5 m x 0.5 m) or 1 m^2, but they can be of any size or shape, even a single point. The quadrats used to sample plant communities are often 0.25 m^2. This size is ideal for low-growing vegetation, but quadrat size needs to be adjusted to habitat type. The quadrat must be large enough to be representative of the community, but not so large as to take a very long time to use.

A quadrat covering an area of 0.25 m^2 is suitable for most low growing plant communities, such as this alpine meadow, fields, and grasslands.

Larger quadrats (e.g. 1 m^2) are needed for communities with shrubs and trees. Quadrats as large as 4 m x 4 m may be needed in woodlands.

Small quadrats (0.01 m^2 or 100 mm x 100 mm) are appropriate for lichens and mosses on rock faces and tree trunks.

How Many Quadrats?

As well as deciding on a suitable quadrat size, the other consideration is how many quadrats to take (the sample size). In species-poor or very homogeneous habitats, a small number of quadrats will be sufficient. In species-rich or heterogeneous habitats, more quadrats will be needed to ensure that all species are represented adequately.

Determining the number of quadrats needed

- Plot the cumulative number of species recorded (on the *y* axis) against the number of quadrats already taken (on the *x* axis).
- The point at which the curve levels off indicates the suitable number of quadrats required.

Fewer quadrats are needed in species-poor or very uniform habitats, such as this bluebell woodland.

Describing Vegetation

Density (number of individuals per unit area) is a useful measure of abundance for animal populations, but can be problematic in plant communities where it can be difficult to determine where one plant ends and another begins. For this reason, plant abundance is often assessed using **percentage cover**. Here, the percentage of each quadrat covered by each species is recorded, either as a numerical value or using an abundance scale such as the ACFOR scale.

The ACFOR Abundance Scale

A = Abundant (30% +)

C = Common (20-29%)

F = Frequent (10-19%)

O = Occasional (5-9%)

R = Rare (1-4%)

The AFCOR scale could be used to assess the abundance of species in this wildflower meadow. Abundance scales are subjective, but it is not difficult to determine which abundance category each species falls into.

1. Describe one difference between the methods used to assess species abundance in plant and in animal communities:

2. Identify the main consideration when determining appropriate quadrat size:

3. Identify the main consideration when determining number of quadrats:

4. Explain two main disadvantages of using the ACFOR abundance scale to record information about a plant community:

    (a)

    (b)

Sampling a Leaf Litter Population

The diagram on the following page represents an area of leaf litter from a forest floor with a resident population of organisms. The distribution of four animal species as well as the arrangement of leaf litter is illustrated. Leaf litter comprises leaves and debris that have dropped off trees to form a layer of detritus. This exercise is designed to practice the steps required in planning and carrying out a sampling of a natural population. It is desirable, but not essential, that students work in groups of 2–4.

1. **Decide on the sampling method**
   For the purpose of this exercise, it has been decided that the populations to be investigated are too large to be counted directly and a quadrat sampling method is to be used to estimate the average density of the four animal species as well as that of the leaf litter.

2. **Mark out a grid pattern**
   Use a ruler to mark out 3 cm intervals along each side of the sampling area (area of quadrat = 0.03 x 0.03 m). **Draw lines** between these marks to create a 6 x 6 grid pattern (total area = 0.18 x 0.18 m). This will provide a total of 36 quadrats that can be investigated.

3. **Number the axes of the grid**
   Only a small proportion of the possible quadrat positions are going to be sampled. It is necessary to select the quadrats in a random manner. It is not sufficient to simply guess or choose your own on a 'gut feeling'. The best way to choose the quadrats randomly is to create a numbering system for the grid pattern and then select the quadrats from a random number table. Starting at the *top left hand corner*, **number the columns** and **rows** from 1 to 6 on each axis.

4. **Choose quadrats randomly**
   To select the required number of quadrats randomly, use random numbers from a random number table. The random numbers are used as an index to the grid coordinates. Choose 6 quadrats from the total of 36 using table of random numbers provided for you at the bottom of the opposite page. Make a note of which column of random numbers you choose. Each member of your group should choose a different set of random numbers (i.e. different column: A–D) so that you can compare the effectiveness of the sampling method.

   Column of random numbers chosen: _____

   NOTE: Highlight the boundary of each selected quadrat with coloured pen/highlighter.

5. **Decide on the counting criteria**
   Before the counting of the individuals for each species is carried out, the criteria for counting need to be established.

   There may be some problems here. You must decide before sampling begins as to what to do about individuals that are only partly inside the quadrat. Possible answers include:
   (a) Only counting individuals if they are completely inside the quadrat.
   (b) Only counting individuals that have a clearly defined part of their body inside the quadrat (such as the head).
   (c) Allowing for 'half individuals' in the data (e.g. 3.5 snails).
   (d) Counting an individual that is inside the quadrat by half or more as one complete individual.

   Discuss the merits and problems of the suggestions above with other members of the class (or group). You may even have counting criteria of your own. Think about other factors that could cause problems with your counting.

6. **Carry out the sampling**
   Carefully examine each selected quadrat and **count the number of individuals** of each species present. Record your data in the spaces provided on the opposite page.

7. **Calculate the population density**
   Use the combined data TOTALS for the sampled quadrats to estimate the average density for each species by using the formula:

   $$\text{Density} = \frac{\text{Total number in all quadrats sampled}}{\text{Number of quadrats sampled} \times \text{area of a quadrat}}$$

   Remember that a total of 6 quadrats are sampled and each has an area of 0.0009 m². The density should be expressed as the number of individuals *per square meter* (no. m^{-2}).

   Woodlouse: [ ]   False scorpion: [ ]
   Centipede: [ ]   Leaf: [ ]
   Springtail: [ ]

8. (a) In this example the animals are not moving. Describe the problems associated with sampling moving organisms. Explain how you would cope with sampling these same animals if they were really alive and very active:

   _____
   _____
   _____

   (b) Carry out a direct count of all 4 animal species and the leaf litter for the whole sample area (all 36 quadrats). Apply the data from your direct count to the equation given in (7) above to calculate the actual population density (remember that the number of quadrats in this case = 36):

   Woodlouse: [ ]   Centipede: [ ]   False scorpion: [ ]   Springtail: [ ]   Leaf: [ ]

   Compare your estimated population density to the actual population density for each species:

   _____
   _____
   _____

Related activities: Quadrat Sampling, Quadrat-Based Estimates

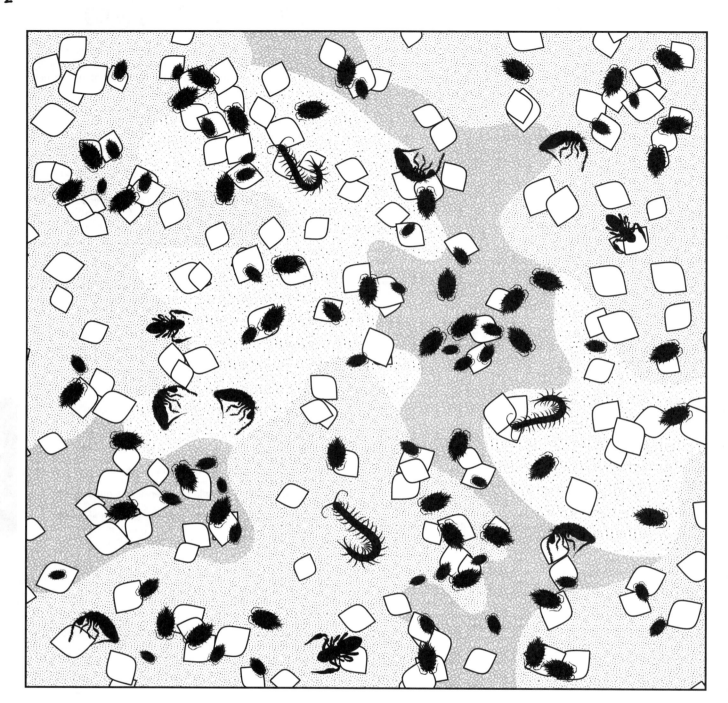

Transect Sampling

A **transect** is a line placed across a community of organisms. Transects are usually carried out to provide information on the **distribution** of species in the community. This is of particular value in situations where environmental factors change over the sampled distance. This change is called an **environmental gradient** (e.g. up a mountain or across a seashore). The usual practice for small transects is to stretch a string between two markers. The string is marked off in measured distance intervals, and the species at each marked point are noted. The sampling points along the transect may also be used for the siting of quadrats, so that changes in density and community composition can be recorded. Belt transects are essentially a form of continuous quadrat sampling. They provide more information on community composition but can be difficult to carry out. Some transects provide information on the vertical, as well as horizontal, distribution of species (e.g. tree canopies in a forest).

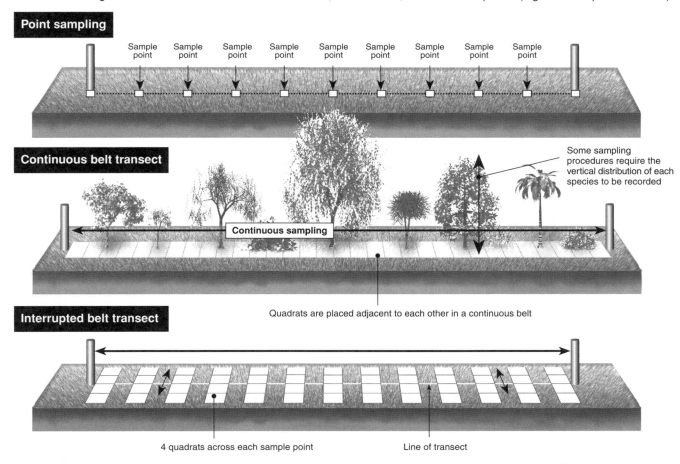

1. Belt transect sampling uses quadrats placed along a line at marked intervals. In contrast, point sampling transects record only the species that are touched or covered by the line at the marked points.

   (a) Describe one disadvantage of belt transects: _____

   (b) Explain why line transects may give an unrealistic sample of the community in question: _____

   (c) Explain how belt transects overcome this problem: _____

   (d) Describe a situation where the use of transects to sample the community would be inappropriate: _____

2. Explain how you could test whether or not a transect sampling interval was sufficient to accurately sample a community: _____

Kite graphs are an ideal way in which to present distributional data from a belt transect (e.g. abundance or percentage cover along an environmental gradient. Usually, they involve plots for more than one species. This makes them good for highlighting probable differences in habitat preference between species. Kite graphs may also be used to show changes in distribution with time (e.g. with daily or seasonal cycles).

3. The data on the right were collected from a rocky shore field trip. Periwinkles from four common species of the genus *Littorina* were sampled in a continuous belt transect from the low water mark, to a height of 10 m above that level. The number of each of the four species in a 1 m² quadrat was recorded.

Plot a **kite graph** of the data for all four species on the grid below. Be sure to choose a scale that takes account of the maximum number found at any one point and allows you to include all the species on the one plot. Include the scale on the diagram so that the number at each point on the kite can be calculated.

Field data notebook
Numbers of periwinkles (4 common species) showing vertical distribution on a rocky shore

Height above low water (m)	L. littorea	L. saxatalis	L. neritoides	L. littoralis
0-1	0	0	0	0
1-2	1	0	0	3
2-3	3	0	0	17
3-4	9	3	0	12
4-5	15	12	0	1
5-6	5	24	0	0
6-7	2	9	2	0
7-8	0	2	11	0
8-9	0	0	47	0
9-10	0	0	59	0

Sampling Animal Populations

Unlike plants, most animals are highly mobile and present special challenges in terms of sampling them **quantitatively** to estimate their distribution and abundance. The equipment available for sampling animals ranges from various types of nets and traps (below), to more complex electronic devices, such as those used for radio-tracking large mobile species.

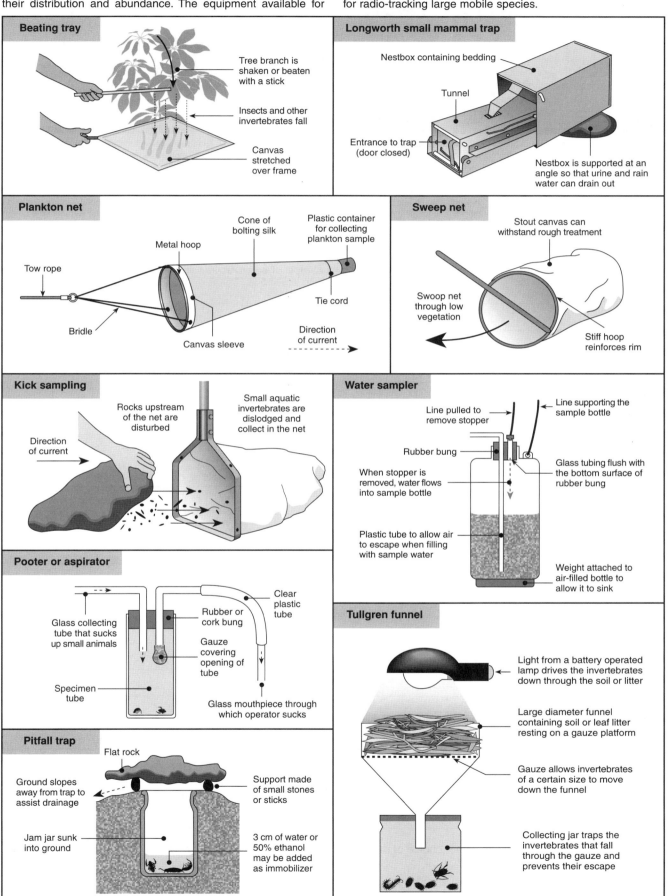

Electrofishing a stream (Sweden)

Adelie penguins fitted with transmitters

Electronic bat detector showing frequency dial

Electrofishing: An effective, but expensive method of sampling larger stream animals (e.g. fish). Wearing a portable battery backpack, the operator walks upstream holding the anode probe and a net. The electrical circuit created by the anode and the stream bed stuns the animals, which are netted and placed in a bucket to recover. After analysis (measurement, species, weights) the animals are released.

Radio-tracking: A relatively non-invasive method of examining many features of animal populations, including movement, distribution, and habitat use. A small transmitter with an antenna (arrowed) is attached to the animal. The transmitter emits a pulsed signal which is picked up by a receiver. In difficult terrain, a tracking antenna can be used in conjunction with the receiver to accurately fix an animal's position.

Electronic detection devices: To sample nocturnal, highly mobile species such as bats, electronic devices, such as the bat detector illustrated above, can be used to estimate population density. In this case, the detector is tuned to the particular frequency of the hunting clicks emitted by the bat species of interest. The number of calls recorded per unit time can be used to estimate numbers within a certain area.

1. Describe what each the following types of sampling equipment is used for in a sampling context:

    (a) Kick sampling technique: _Provides a semi-quantitative sample of substrate-dwelling stream invertebrates_

    (b) Beating tray: _____

    (c) Longworth small mammal trap: _____

    (d) Plankton net: _____

    (e) Sweep net: _____

    (f) Water sampler: _____

    (g) Pooter: _____

    (h) Tullgren funnel: _____

    (i) Pitfall trap: _____

2. Explain why pitfall traps are not recommended for estimates of population density: _____

3. (a) Explain what influence mesh size might have on the sampling efficiency of a plankton net: _____

    (b) Explain how this would affect your choice of mesh size when sampling animals in a pond: _____

Mark and Recapture Sampling

The mark and recapture method of estimating population size is used in the study of animal populations where individuals are highly mobile. It is of no value where animals do not move or move very little. The number of animals caught in each sample must be large enough to be valid. The technique is outlined in the diagram below.

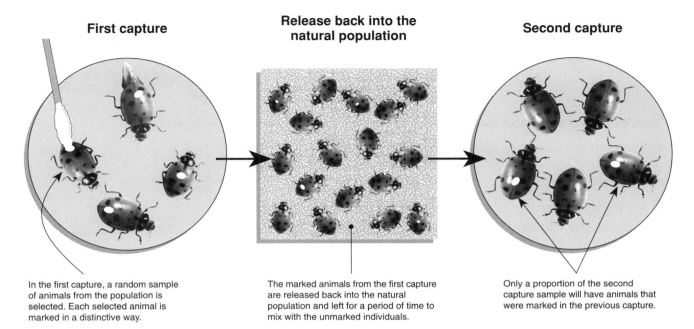

First capture: In the first capture, a random sample of animals from the population is selected. Each selected animal is marked in a distinctive way.

Release back into the natural population: The marked animals from the first capture are released back into the natural population and left for a period of time to mix with the unmarked individuals.

Second capture: Only a proportion of the second capture sample will have animals that were marked in the previous capture.

The Lincoln Index

$$\text{Total population} = \frac{\text{No. of animals in 1st sample (all marked)} \times \text{Total no. of animals in 2nd sample}}{\text{Number of marked animals in the second sample (recaptured)}}$$

The mark and recapture technique comprises a number of simple steps:

1. The population is sampled by capturing as many of the individuals as possible and practical.
2. Each animal is marked in a way to distinguish it from unmarked animals (unique mark for each individual not required).
3. Return the animals to their habitat and leave them for a long enough period for complete mixing with the rest of the population to take place.
4. Take another sample of the population (this does not need to be the same sample size as the first sample, but it does have to be large enough to be valid).
5. Determine the numbers of marked to unmarked animals in this second sample. Use the equation above to estimate the size of the overall population.

1. For this exercise you will need several boxes of matches and a pen. Work in a group of 2-3 students to 'sample' the population of matches in the full box by using the mark and recapture method. Each match will represent one animal.

    (a) Take out 10 matches from the box and mark them on 4 sides with a pen so that you will be able to recognize them from the other unmarked matches later.
    (b) Return the marked matches to the box and shake the box to mix the matches.
    (c) Take a sample of 20 matches from the same box and record the number of marked matches and unmarked matches.
    (d) Determine the total population size by using the equation above.
    (e) Repeat the sampling 4 more times (steps b–d above) and record your results:

	Sample 1	Sample 2	Sample 3	Sample 4	Sample 5
Estimated population					

    (f) Count the actual number of matches in the matchbox : _____

    (g) Compare the actual number to your estimates. By how much does it differ: _____

2. In 1919 a researcher by the name of Dahl wanted to estimate the number of trout in a Norwegian lake. The trout were subject to fishing so it was important to know how big the population was in order to manage the fish stock. He captured and marked 109 trout in his first sample. A few days later, he caught 177 trout in his second sample, of which 57 were marked. Use the **Lincoln index** (on the opposite page) to estimate the total population size:

Size of 1st sample: _____

Size of 2nd sample: _____

No. marked in 2nd sample: _____

Estimated total population: _____

3. Describe some of the problems with the mark and recapture method if the second sampling is:

   (a) Left too long a time before being repeated: _____

   (b) Too soon after the first sampling: _____

4. Describe two important assumptions being made in this method of sampling, that would cause the method to fail if they were not true:

   (a) _____

   (b) _____

5. Some types of animal would be unsuitable for this method of population estimation (i.e. the method would not work).

   (a) Name an animal for which this method of sampling would not be effective: _____

   (b) Explain your answer above: _____

6. Describe three methods for marking animals for mark and recapture sampling. Take into account the possibility of animals shedding their skin, or being difficult to get close to again:

   (a) _____

   (b) _____

   (c) _____

7. Scientists in the UK and Canada have, at various times since the 1950s, been involved in computerized tagging programs for Northern cod (a species once abundant in Northern Hemisphere waters but now severely depleted). Describe the type of information that could be obtained through such tagging programs: _____

Indirect Sampling

79

If populations are small and easily recognized they may be monitored directly quite easily. However, direct measurement of elusive, or widely dispersed populations is not always feasible. In these cases, indirect methods can be used to assess population abundance, provide information on habitat use and range, and enable biologists to link habitat quality to species presence or absence. Indirect sampling methods provide less reliable measures of abundance than direct sampling methods, such as mark and recapture, but are widely used nevertheless. They rely on recording the signs of a species, e.g. scat, calls, tracks, and rubbings or markings on vegetation, and using these to assess population abundance. In Australia, the Environmental Protection Agency (EPA) provides a Frog Census Datasheet (below) on which volunteers record details about frog populations and habitat quality in their area. This program enables the EPA to gather information across Australia. Another example of an alternative method of population sampling used in New Zealand is the Kiwi Recovery Program (see following page).

Recording a date and accurate map reference is important

Population estimates are based on the number of frog calls recorded by the observer

To sample nocturnal, highly mobile species, e.g. bats, electronic devices, such as the bat detector above, can be used to estimate population density. In this case, the detector is tuned to the particular frequency of the hunting clicks emitted by specific bat species. The number of calls recorded per unit time can be used to estimate numbers per area.

The analysis of animal tracks allows wildlife biologists to identify habitats in which animals live and to conduct population surveys. Interpreting tracks accurately requires considerable skill as tracks may vary in appearance even when from the same individual. Tracks are particularly useful as a way to determine habitat use and preference.

All animals leave scats (feces) which are species specific and readily identifiable. Scats can be a valuable tool by which to gather data from elusive, nocturnal, easily disturbed, or highly mobile species. Fecal analyses can provide information on diet, movements, population density, sex ratios, age structure, and even genetic diversity.

1. Describe two kinds of indirect signs that could be used to detect the presence of frogs:

    (a) _____ (b) _____

2. (a) Describe the kind of information that the EPA would gather from their Frog Census Datasheet: _____

    _____

    _____

    _____

    (b) Explain a use for this information: _____

    _____

© Biozone International 2006-2007
Photocopying Prohibited

RA 1

The **Kiwi Reporting Card** is issued to trampers and conservation groups who are helping New Zealand's Department of Conservation to gather census data on kiwi. Read all parts of the card carefully and answer the questions below.

KIWI REPORTING CARD

Please complete this form as fully as possible and return to the Department of Conservation.

KIWI RECOVERY

DATE:	OBSERVER'S DETAILS:
LOCATION:	NAME
	ADDRESS
MAP SERIES SHEET GRID REFERENCE	PHONE Home/Work
NUMBER OF KIWI SEEN (Other, please specify)	COMMENTS (Vegetation, habitat, dogs or other predators seen)
NUMBER OF KIWI HEARD — Male calls / Female calls	SIGNS OF KIWI PRESENT (eg. footprints, probeholes)

Are you 100% sure that what you saw or heard was a kiwi? YES/NO

NOTES
1. The call of the male kiwi is a repetitive (8-25 notes) high-pitched whistle.
2. The call of the female kiwi is a repetitive (10-20 notes) coarse rasping note.
3. Weka, moreporks and possums are often confused with kiwi calls.
4. Footprints are about the size of a domestic chicken and are often found in mud or snow.
5. Probeholes usually occur in groups and look like a screwdriver has been pushed into the ground, rotated and pulled out again. They are about 10cm deep.
6. This form may also be used to report other species of wildlife such as kaka, kokako, blue duck, bats, etc. Please ensure you clearly identify what you are recording.
7. Post this card to the Department of Conservation (address provided on the back) or leave it with a hut warden.

3. Describe three kinds of indirect signs that can be used to detect the presence of kiwi:

   (a) _____

   (b) _____

   (c) _____

4. Explain why it is not easy to carry out a direct count of a kiwi population:

5. Describe the attempts that the organizers of this data collecting program have made to ensure that the people recording their observations are correctly identifying kiwi signs:

6. Explain why the comments section on the card requests information on the habitat, dogs, or other predators seen:

7. Describe one other indirect method of population sampling and outline its advantages and drawbacks:

© Biozone International 2006-2007
Photocopying Prohibited

Sampling Using Radio-tracking

Field work involving difficult terrain, aquatic environments, or highly mobile, secretive, or easily disturbed species, has been greatly assisted in recent years by the use of radio-transmitter technology. Radio-tracking is particularly suited to population studies of threatened species (because it is relatively non-invasive) and of pests (because their dispersal and habitat use can be monitored). There are many reasons why radio-tracking is used to follow animal movements. Importantly, radio-tracking can be used to quickly obtain accurate information about an animal's home range. This knowledge is essential to understanding dispersal, distribution, habitat use, and competitive relationships. The information can be used to manage an endangered species effectively, or to plan efficient pest control operations. Satellite transmitters can be used to study the large scale migratory movements of large animals and marine species, which are more difficult to follow using the usual VHF radio-tracking equipment.

Adelie penguins with transmitters

Scanning receiver with 400 channels

Aerial tracking of rooks (a pest bird)

Hand-held portable antenna

A transmitter emits a **pulsed signal** that can be picked up by a **receiver**. Transmitters may be short or long range. Each transmitter has an antenna which may be whiplike (above left) or a loop type, which is often incorporated into a collar. Transmitter size and antenna length is set so that there is no interference with the animal's behavior. Usually the weight of the transmitter does not exceed 2% of the animal's body weight. Receivers pick up the transmitter signal and different channels can be used to display signals from many transmitters. Many are very portable and are used in difficult terrain.

A tracking antenna together with the receiver can be used to home in on an animal. An antenna is directional and so can accurately fix an animal's position. Antennae can be mounted onto light aircraft or off-road vehicles to provide mobile tracking over large areas. For work in inaccessible or difficult terrain, portable, hand-held antennae are used. A folding model is useful in thick scrub and is easily carried when not in use. The more portable the antenna the less powerful it is in terms of picking up the transmitter signal. Very portable antennae are useful only for short range tracking.

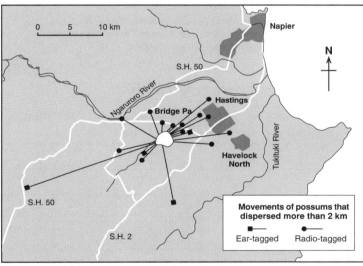

A possum (a major pest in New Zealand) with a transmitter around its neck. The antenna can be seen above the back. Radio-tracking is used on pest species to determine dispersal rates, distribution and habitat use. With this information, pest control can be implemented more effectively. The map to the left shows the dispersal movements of radio-tagged and ear-tagged possums on the east coast of New Zealand.

Photos courtesy of Sirtrack Ltd except where indicated otherwise

Giant weta with transmitter

Long tailed bat with transmitter

Male tuatara wearing transmitter

Goldstripe gecko with transmitter

Radio-tracking equipment is now widely used in many aspects of conservation work. Very small animals, such as weta (left), bats, and small lizards require very small transmitters, weighing less than 1 g. These are often glued on, or mounted on a body harness. Radio-tracking has been used to study the movements and habitat use of animals as diverse as large insects, snails, birds, and bats. Such knowledge allows conservation organizations to develop better management strategies for these species in the wild.

Tracking devices are used extensively by conservation organizations and university research groups to study reptiles. The findings of radio-tracking studies of tuatara in New Zealand indicated that they will use artificial burrows after translocation to islands. The studies also showed and that captive reared individuals have different dispersal patterns to those moved directly from the wild. The niche requirements of the goldstripe gecko were investigated in a similar way, using short range transmitters, located using harmonic radar.

TRANSMITTER SENSING OPTIONS

Transmitters can be set to provide specific information about the activity or physiological state of an animal. Sensing options include:

Activity sensing
The transmitter records changes in posture or behavior.

Mortality transmitter
Transmitter pulse rate will double (or halve) if it is not moved for a certain time (indicating death).

Heart rate monitors
The transmitter emits a pulse every time the animal's heart beats. It has particular application in studies of animal responses and stress.

Temperature sensing
In these systems, the pulse rate of the transmitter varies according to animal temperature. A calibration curve or a decoding unit on the receiver is used to convert the pulse rate information to a temperature.

Audio transmitters
These pick up animal sounds. Direct observation of the animal is sometimes needed to interpret the sound.

Accurate tracking of movements within a **home range** provides information about habitat use and range requirements. This allows species (such as this New Zealand native rat) to be monitored and managed within their own habitats.

A range of different tracking collars

Tracking migratory movements provides the opportunity to manage and protect threatened species (such as this green turtle) during all phases of their migration.

1. Describe two applications of radio-tracking technology to the management of endangered species:

   (a) _____

   (b) _____

2. Discuss the advantages and disadvantages of using radio-tracking as a sampling technique: _____

3. Explain why radio-tracking is often used for monitoring pest species for which elimination (not conservation) is the goal:

4. Identify one major constraint of transmitter design and explain why this needs to be considered:

Classification of Organisms

Identifying and describing the diversity of organisms

Five kingdom classification, new classification schemes, binomial nomenclature, identifying plant and animal diversity. Taxonomic keys.

Learning Objectives

☐ 1. Compile your own glossary from the **KEY WORDS** displayed in **bold type** in the learning objectives below.

Classification Systems

The five kingdoms *(pages 84, 86-92, 96-105)*

☐ 2. Describe the principles and importance of scientific classification. Recognise **taxonomy** as the study of the theory and practice of classification.

☐ 3. Describe the **distinguishing features** of each kingdom in the **five kingdom classification system**:
  • **Prokaryotae** (Monera): bacteria and cyanobacteria.
  • **Protista**: includes the algae and protozoans.
  • **Fungi**: includes yeasts, moulds, and mushrooms.
  • **Plantae**: includes mosses, liverworts, tracheophytes.
  • **Animalia**: all invertebrate phyla and the chordates.

☐ 4. Demonstrate a working knowledge of taxonomy by classifying familiar organisms. Recognise at least seven major **taxonomic categories**: **kingdom**, **phylum**, **class**, **order**, **family**, **genus**, and **species**.

☐ 5. Appreciate that taxonomic categories should not be confused with **taxa** (sing. **taxon**), which are groups of real organisms: "genus" is a taxonomic category, whereas the genus *Drosophila* is a taxon.

☐ 6. Understand the basis for assigning organisms to different taxonomic categories. Recall what is meant by a **distinguishing feature**. Appreciate that species are classified on the basis of **shared derived characters** rather than primitive (ancestral) characters. *For example, within the vertebrates, the presence of a backbone is a derived, therefore a distinguishing, feature. Within the mammals, the backbone is an ancestral feature and is not distinguishing, whereas mammary glands (a distinguishing feature) are derived.*

☐ 7. Explain how **binomial nomenclature** is used to classify organisms. Appreciate the problems associated with using **common names** to describe organisms.

☐ 8. Explain the relationship between classification and phylogeny. Appreciate that newer classification schemes attempt to better reflect the **phylogeny** of organisms.

New classification schemes *(pages 84-85)*

☐ 9. Recognise the recent reclassification of organisms into three **domains**: **Archaea**, **Eubacteria**, and **Eukarya**. Explain the basis and rationale for this classification.

☐ 10. Appreciate that **cladistics** provides a method of classification based on relatedness, and that it emphasizes the presence of **shared derived characters**. Discuss the benefits and disadvantages associated with cladistic schemes.

Classification keys *(pages 93-95)*

☐ 11. Explain what a **classification key** is and what it is used for. Describe the essential features of a classification key. Use a simple taxonomic key to recognise and classify some common organisms.

See page 7 for details of publishers of periodicals:

STUDENT'S REFERENCE

■ **How Many Mammals in Britain?** Biol. Sci. Rev., 12(4) March 2000, pp. 18-22. *This article examines the nature of species definitions and the diversity and abundance of mammals in Britain.*

■ **Taxonomy: The Naming Game Revisited** Biol. Sci. Rev., 9(5) May 1997, pp. 31-35. *New tools for taxonomy and how they are used (includes the exemplar of the reclassification of the kingdoms).*

■ **The Living Dead** New Scientist, 13 October 2001 (Inside Science). *An account of the nature of viruses; their classification and diversity, as well as their evolution and their role as pathogens.*

■ **A Passion for Order** National Geographic, 211(6) June 2007, pp. 73-87. *The history of Carl Linnaeus and the classification of plant species.*

■ **The Family Line - The Human-Cat Connection** National Geographic, 191(6) June 1997, pp. 77-85. *An examination of the genetic diversity and lineages within the felidae. A good context within which to study classification.*

■ **Biological Taxonomy** Biol. Sci. Rev., 8(3) Jan. 1996, pp. 34-37. *Taxonomy, including a clear, concise treatment of cladistics and phenetics.*

TEACHER'S REFERENCE

■ **What's in a Name?** Scientific American, Nov. 2004, pp. 20-21. *A proposed classification system called phyloclode, based solely on phylogeny.*

■ **Family Feuds** New Scientist, 24 January 1998, pp. 36-40. *Molecular and morphological analysis used for determining species interrelatedness.*

■ **The Loves of the Plants** Scientific American, Feb. 1996, pp. 98-103. *The classification of plants and the development of keys to plant identification.*

■ **Computers, DNA, & Evolution** Biol. Sci. Rev., 11(5) May 1999, pp. 24-29. *Using DNA sequence analysis to study evolutionary relationships.*

■ **Gladiators: A New Order of Insects** Scientific American, Nov. 2002, pp. 42-47. *This account describes the characteristics, distribution, life history, and taxonomy of a new order of insects.*

■ **Extremophiles** Scientific American, April 1997, pp. 66-71. *The biology and taxonomy of the microbial populations of extreme environments.*

■ **Is it Kingdoms or Domains?** American Biology Teacher, 66(4), April 2004, pp. 268-276. *How many kingdoms are there, which ones should we recognize, and how do domains fit into a tradition of kingdom-based classification?*

See pages 4-5 for details of how to access **Bio Links** from our web site: **www.thebiozone.com** From Bio Links, access sites under the topics:
GENERAL BIOLOGY ONLINE RESOURCES
Glossaries: • Taxonomy glossary ... *and others*

MICROBIOLOGY > **General Microbiology**: • Bacteriology homepage • Biological identity of the prokaryotes • British Mycological Society • Major groups of prokaryotes • The microbe zoo • Microbiology webbed out .. *and others*

PLANT BIOLOGY: Classification and Diversity: • Flowering plant diversity • Natural perspective: Plant Kingdom • Introduction to the Plantae • Vascular plant families ... *and others*

Presentation MEDIA to support this topic:
ECOLOGY:
  • Biodiversity and Conservation

The New Tree of Life

With the advent of more efficient genetic (DNA) sequencing technology, the genomes of many bacteria began to be sequenced. In 1996, the results of a scientific collaboration examining DNA evidence confirmed the proposal that life comprises three major evolutionary lineages (domains) and not two as was the convention. The recognized lineages were the **Eubacteria**, the **Eukarya** and the **Archaea** (formerly the Archaebacteria). The new classification reflects the fact that there are very large differences between the archaea and the eubacteria. All three domains probably had a distant common ancestor.

A Five (or Six) Kingdom World (right)

The diagram (right) represents the **five kingdom system** of classification commonly represented in many biology texts. It recognizes two basic cell types: prokaryote and eukaryote. The domain Prokaryota includes all bacteria and cyanobacteria. Domain Eukaryota includes protists, fungi, plants, and animals. More recently, based on 16S ribosomal RNA sequence comparisons, Carl Woese divided the prokaryotes into two kingdoms, the Eubacteria and Archaebacteria. Such **six-kingdom systems** are also commonly recognized in texts.

A New View of the World (below)

In 1996, scientists deciphered the full DNA sequence of an unusual bacterium called *Methanococcus jannaschii*. An **extremophile**, this methane-producing archaebacterium lives at 85°C; a temperature lethal for most bacteria as well as eukaryotes. The DNA sequence confirmed that life consists of three major evolutionary lineages, not the two that have been routinely described. Only 44% of this archaebacterium's genes resemble those in bacteria or eukaryotes, or both.

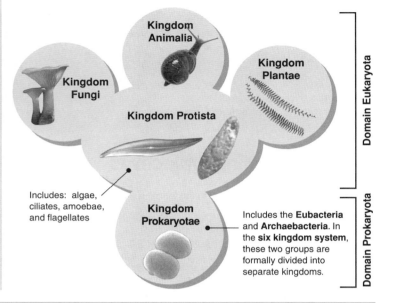

Domain Eubacteria
Lack a distinct nucleus and cell organelles. Generally prefer less extreme environments than Archaea. Includes well-known pathogens, many harmless and beneficial species, and the cyanobacteria (photosynthetic bacteria containing the pigments chlorophyll *a* and phycocyanin).

Domain Archaea
Closely resemble eubacteria in many ways but cell wall composition and aspects of metabolism are very different. Live in extreme environments similar to those on primeval Earth. They may utilize sulfur, methane, or halogens (chlorine, fluorine), and many tolerate extremes of temperature, salinity, or pH.

Domain Eukarya
Complex cell structure with organelles and nucleus. This group contains four of the kingdoms classified under the more traditional system. Note that Kingdom Protista is separated into distinct groups: e.g. amoebae, ciliates, flagellates.

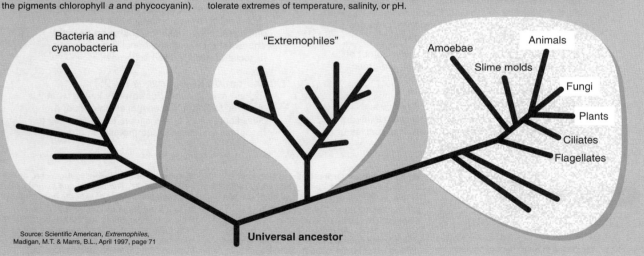

Source: Scientific American, *Extremophiles*, Madigan, M.T. & Marrs, B.L., April 1997, page 71

1. Explain why some scientists have recommended that the conventional classification of life be revised so that the Archaea, Eubacteria and Eukarya are three separate domains:

2. Describe two features of the new classification scheme that are very different from the five kingdom classification:

   (a)

   (b)

Related activities: New Classification Schemes, Features of Taxonomic Groups, Features of the Five Kingdoms

New Classification Schemes

Taxonomy is the study of classification. Ever since Darwin, the aim of classification has been to organise species, and to reflect their evolutionary history (**phylogeny**). Each successive group in the taxonomic hierarchy should represent finer and finer branching from a common ancestor. In order to reconstruct evolutionary history, phylogenetic trees must be based on features that are due to shared ancestry (homologies). Traditional taxonomy has relied mainly on **morphological characters** to do this. Modern technology has assisted taxonomy by providing **biochemical evidence** (from proteins and DNA) for the relatedness of species. The most familiar approach to classifying organisms is to use **classical evolutionary taxonomy**. It considers branching sequences and overall likeness. A more recent approach has been to use **cladistics**: a technique which emphasizes phylogeny or relatedness, usually based on biochemical evidence (and largely ignoring morphology or appearance). Each branch on the tree marks the point where a new species has arisen by evolution. Traditional and cladistic schemes do not necessarily conflict, but there have been reclassifications of some taxa (notably the primates, but also the reptiles, dinosaurs, and birds). Traditional taxonomists criticise cladistic schemes because they do not recognise the amount of visible change in morphology that occurs in species after their divergence from a common ancestor. Popular classifications will probably continue to reflect similarities and differences in appearance, rather than a strict evolutionary history. In this respect, they are a compromise between phylogeny and the need for a convenient filing system for species diversity.

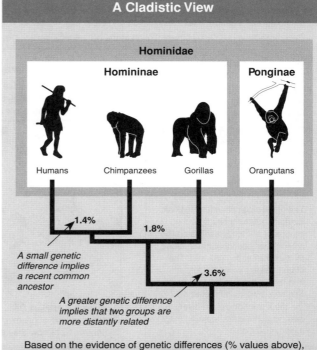

A Classical Taxonomic View

On the basis of overall anatomical similarity (e.g. bones and limb length, teeth, musculature), apes are grouped into a family (Pongidae) that is separate from humans and their immediate ancestors (Hominidae). The family Pongidae (the great apes) is not monophyletic (of one phylogeny), because it stems from an ancestor that also gave rise to a species in another family (i.e. humans). This traditional classification scheme is now at odds with schemes derived after considering genetic evidence.

A Cladistic View

Based on the evidence of genetic differences (% values above), chimpanzees and gorillas are more closely related to humans than to orangutans, and chimpanzees are more closely related to humans than they are to gorillas. Under this scheme there is no true family of great apes. The family Hominidae includes two subfamilies: Ponginae and Homininae (humans, chimpanzees, and gorillas). This classification is monophyletic: the Hominidae includes all the species that arise from a common ancestor.

1. Briefly explain the benefits of classification schemes based on:

   (a) Morphological characters: _____

   (b) Relatedness in time (from biochemical evidence): _____

2. Describe the contribution of biochemical evidence to taxonomy: _____

3. Based on the diagram above, state the family to which the chimpanzees belong under:

   (a) A traditional scheme: _____   (b) A cladistic scheme: _____

Features of Taxonomic Groups

In order to distinguish organisms, it is desirable to classify and name them (a science known as **taxonomy**). An effective classification system requires features that are distinctive to a particular group of organisms. Revised classification systems, recognizing three domains (rather than five or six kingdoms) are now recognized as better representations of the true diversity of life. However, for the purposes of describing the groups with which we are most familiar, the five kingdom system (used here) is still appropriate. The distinguishing features of some major **taxa** are provided in the following pages by means of diagrams and brief summaries. Note that most animals show **bilateral symmetry** (body divisible into two halves that are mirror images). **Radial symmetry** (body divisible into equal halves through various planes) is a characteristic of cnidarians and ctenophores.

Kingdom: PROKARYOTAE (Bacteria)

- Also known as monerans or prokaryotes.
- Two major bacterial lineages are recognized: the primitive **Archaebacteria** and the more advanced **Eubacteria**.
- All have a prokaryotic cell structure: they lack the nuclei and chromosomes of eukaryotic cells, and have smaller (70S) ribosomes.
- Have a tendency to spread genetic elements across species barriers by sexual conjugation, viral transduction and other processes.
- Can reproduce rapidly by binary fission in the absence of sex.
- Have evolved a wider variety of metabolism types than eukaryotes.
- Bacteria grow and divide or aggregate into filaments or colonies of various shapes.
- They are taxonomically identified by their appearance (form) and through biochemical differences.

Species diversity: 10 000 + Bacteria are rather difficult to classify to the species level because of their relatively rampant genetic exchange, and because their reproduction is usually asexual.

Eubacteria

- Also known as 'true bacteria', they probably evolved from the more ancient Archaebacteria.
- Distinguished from Archaebacteria by differences in cell wall composition, nucleotide structure, and ribosome shape.
- Very diverse group comprises most bacteria.
- The **gram stain** provides the basis for distinguishing two broad groups of bacteria. It relies on the presence of peptidoglycan (unique to bacteria) in the cell wall. The stain is easily washed from the thin peptidoglycan layer of gram negative walls but is retained by the thick peptidoglycan layer of gram positive cells, staining them a dark violet color.

Gram-Positive Bacteria

The walls of gram positive bacteria consist of many layers of peptidoglycan forming a thick, single-layered structure that holds the gram stain.

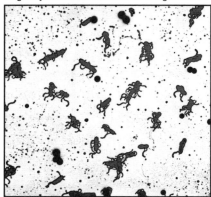

Bacillus alvei: a gram positive, flagellated bacterium. Note how the cells appear dark.

Gram-Negative Bacteria

The cell walls of gram negative bacteria contain only a small proportion of peptidoglycan, so the dark violet stain is not retained by the organisms.

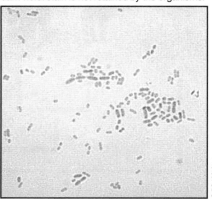

Alcaligenes odorans: a gram negative bacterium. Note how the cells appear pale.

Kingdom: FUNGI

- Heterotrophic.
- Rigid cell wall made of chitin.
- Vary from single celled to large multicellular organisms.
- Mostly saprotrophic (i.e. feeding on dead or decaying material).
- Terrestrial and immobile.

Examples:
Mushrooms/toadstools, yeasts, truffles, morels, molds, and lichens.

Species diversity: 80 000 +

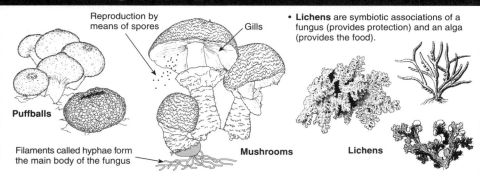

- **Lichens** are symbiotic associations of a fungus (provides protection) and an alga (provides the food).

Puffballs — Filaments called hyphae form the main body of the fungus — Reproduction by means of spores — Gills — Mushrooms — Lichens

Kingdom: PROTISTA

- A diverse group of organisms that do not fit easily into other taxonomic groups.
- Unicellular or simple multicellular.
- Widespread in moist or aquatic environments.

Examples of algae: green, brown, and red algae, dinoflagellates, diatoms.

Examples of protozoa: amoebas, foraminiferans, radiolarians, ciliates.

Species diversity: 55 000 +

Algae 'plant-like' protists

- Autotrophic (photosynthesis)
- Characterized by the type of chlorophyll present

Cell walls of cellulose, sometimes with silica — **Diatom**

Protozoa 'animal-like' protists

- Heterotrophic nutrition and feed via ingestion
- Most are microscopic (5 µm-250 µm)

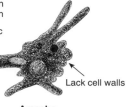

Move via projections called pseudopodia — **Amoeba** — Lack cell walls

Kingdom: PLANTAE

- Multicellular organisms (the majority are photosynthetic and contain chlorophyll).
- Cell walls made of cellulose; Food is stored as starch.
- Subdivided into two major divisions based on tissue structure: **Bryophytes** (non-vascular) and **Tracheophytes** (vascular) plants.

Non-Vascular Plants:
- Non vascular, lacking transport tissues (no xylem or phloem).
- They are small and restricted to moist, terrestrial environments.
- Do not possess 'true' roots, stems or leaves

Phylum Bryophyta: Mosses, liverworts, and hornworts.

Species diversity: 18 600 +

Phylum: Bryophyta

Liverworts — Sexual reproductive structures; Flattened thallus (leaf like structure)

Mosses — Sporophyte: reproduce by spores; Rhizoids anchor the plant into the ground

Vascular Plants:
- Vascular: possess transport tissues.
- Possess true roots, stems, and leaves, as well as stomata.
- Reproduce via spores, not seeds.
- Clearly defined *alternation of sporophyte and gametophyte generations*.

Seedless Plants:
Spore producing plants, includes:
Phylum Filicinophyta: Ferns
Phylum Sphenophyta: Horsetails
Phylum Lycophyta: Club mosses

Species diversity: 13 000 +

Phylum: Lycophyta
Club moss — Leaves

Phylum: Sphenophyta
Horsetail — Leaves

Phylum: Filicinophyta
Fern — Reproduce via spores on the underside of leaf; Large dividing leaves called fronds; Rhizome; Adventitious roots

Seed Plants:
Also called Spermatophyta. Produce seeds housing an embryo. Includes:

Gymnosperms
- Lack enclosed chambers in which seeds develop.
- Produce seeds in cones which are exposed to the environment.

Phylum Cycadophyta: Cycads
Phylum Ginkgophyta: Ginkgoes
Phylum Coniferophyta: Conifers

Species diversity: 730 +

Phylum: Cycadophyta
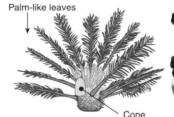
Cycad — Palm-like leaves; Cone

Phylum: Ginkophyta

Ginkgo — Flat leaves

Phylum: Coniferophyta
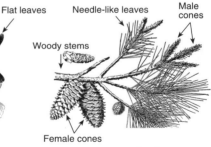
Conifer — Needle-like leaves; Male cones; Woody stems; Female cones

Angiosperms
Phylum: Angiospermophyta
- Seeds in specialized reproductive structures called flowers.
- Female reproductive ovary develops into a fruit.
- Pollination usually via wind or animals.

Species diversity: 260 000 +

The phylum Angiospermophyta may be subdivided into two classes:
Class *Monocotyledoneae* (Monocots)
Class *Dicotyledoneae* (Dicots)

Angiosperms: **Monocotyledons**

Lily
- Flower parts occur in multiples of 3
- Leaves have parallel veins
- Only have one cotyledon (food storage organ)
- Normally herbaceous (non-woody) with no secondary growth

Examples: cereals, lilies, daffodils, palms, grasses.

Angiosperms: **Dicotyledons**
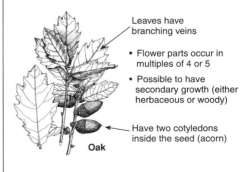
Oak
- Leaves have branching veins
- Flower parts occur in multiples of 4 or 5
- Possible to have secondary growth (either herbaceous or woody)
- Have two cotyledons inside the seed (acorn)

Examples: many annual plants, trees and shrubs.

Kingdom: ANIMALIA

- Over 800 000 species described in 33 existing phyla.
- Multicellular, heterotrophic organisms.
- Animal cells lack cell walls.
- Further subdivided into various major phyla on the basis of body symmetry, type of body cavity, and external and internal structures.

Phylum: Rotifera

- A diverse group of small organisms with sessile, colonial, and planktonic forms.
- Most freshwater, a few marine.
- Typically reproduce via cyclic parthenogenesis.
- Characterized by a wheel of cilia on the head used for feeding and locomotion, a large muscular pharynx (mastax) with jaw like trophi, and a foot with sticky toes.

Species diversity: 1500 +

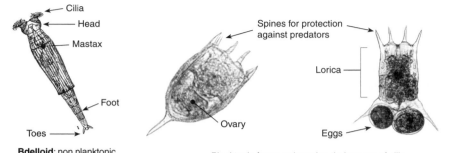

Bdelloid: non planktonic, creeping rotifer

Planktonic forms swim using their crown of cilia

Phylum: Porifera

- Lack organs.
- All are aquatic (mostly marine).
- Asexual reproduction by budding.
- Lack a nervous system.

Examples: sponges.
Species diversity: 8000 +

- Capable of regeneration (the replacement of lost parts)
- Possess spicules (needle-like internal structures) for support and protection

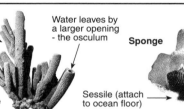

Phylum: Cnidaria

- Two basic body forms:
  Medusa: umbrella shaped and free swimming by pulsating bell.
  Polyp: cylindrical, some are sedentary, others can glide, or somersault or use tentacles as legs.
- Some species have a life cycle that alternates between a polyp stage and a medusa stage.
- All are aquatic (most are marine).

Examples: Jellyfish, sea anemones, hydras, and corals.
Species diversity: 11 000 +

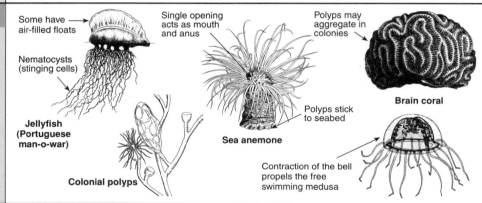

Phylum: Platyhelminthes

- Unsegmented body.
- Flattened body shape.
- Mouth, but no anus.
- Many are parasitic.

Examples: Tapeworms, planarians, flukes.
Species diversity: 20 000 +

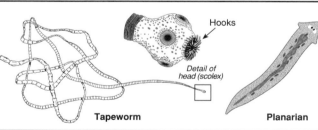

Liver fluke   **Tapeworm**   **Planarian**

Phylum: Nematoda

- Tiny, unsegmented roundworms.
- Many are plant/animal parasites.

Examples: Hookworms, stomach worms, lung worms, filarial worms
Species diversity: 80 000 - 1 million

A general nematode body plan

Phylum: Annelida

- Cylindrical, segmented body with chaetae (bristles).
- Move using hydrostatic skeleton and/or parapodia (appendages).

Examples: Earthworms, leeches, polychaetes (including tubeworms).
Species diversity: 15 000 +

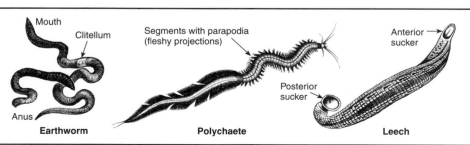

Earthworm   **Polychaete**   **Leech**

Kingdom: ANIMALIA (continued)

Phylum: Mollusca

- Soft bodied and unsegmented.
- Body comprises head, muscular foot, and visceral mass (organs).
- Most have radula (rasping tongue).
- Aquatic and terrestrial species.
- Aquatic species possess gills.

Examples: Snails, mussels, squid.
Species diversity: 110 000 +

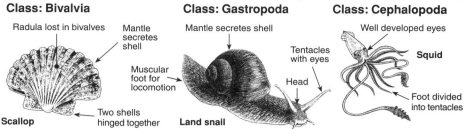

Class: Bivalvia — Scallop (Radula lost in bivalves; Mantle secretes shell; Two shells hinged together)
Class: Gastropoda — Land snail (Mantle secretes shell; Muscular foot for locomotion; Tentacles with eyes; Head)
Class: Cephalopoda — Squid (Well developed eyes; Foot divided into tentacles)

Phylum: Arthropoda

- Exoskeleton made of chitin.
- Grow in stages after moulting.
- Jointed appendages.
- Segmented bodies.
- Heart found on dorsal side of body.
- Open circulation system.
- Most have compound eyes.

Species diversity: 1 million +
Make up 75% of all living animals.

Arthropods are subdivided into the following classes:

Class: Crustacea (crustaceans)
- Mainly marine.
- Exoskeleton impregnated with mineral salts.
- Gills often present.
- Includes: Lobsters, crabs, barnacles, prawns, shrimps, isopods, amphipods
- **Species diversity:** 35 000 +

Class: Arachnida (chelicerates)
- Almost all are terrestrial.
- 2 body parts: cephalothorax and abdomen (except horseshoe crabs).
- Includes: spiders, scorpions, ticks, mites, horseshoe crabs.
- **Species diversity:** 57 000 +

Class: Insecta (insects)
- Mostly terrestrial.
- Most are capable of flight.
- 3 body parts: head, thorax, abdomen.
- Include: Locusts, dragonflies, cockroaches, butterflies, bees, ants, beetles, bugs, flies, and more
- **Species diversity:** 800 000 +

Myriapoda (=many legs)
Class: Diplopoda (millipedes)
- Terrestrial.
- Have a rounded body.
- Eat dead or living plants.
- **Species diversity:** 2000 +

Class: Chilopoda (centipedes)
- Terrestrial.
- Have a flattened body.
- Poison claws for catching prey.
- Feed on insects, worms, and snails.
- **Species diversity:** 7000 +

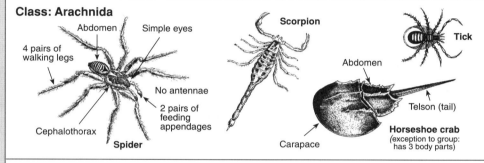

Class: Crustacea — Shrimp (2 pairs of antennae; 3 pairs of mouthparts; Cheliped (first leg); Cephalothorax (fusion of head and thorax); Abdomen; Swimmerets; Walking legs); Crab; Amphipod

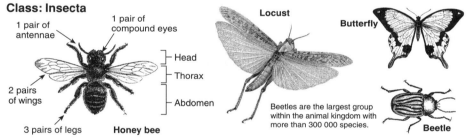

Class: Arachnida — Spider (4 pairs of walking legs; Abdomen; Simple eyes; Cephalothorax; No antennae; 2 pairs of feeding appendages); Scorpion; Tick; Horseshoe crab (Carapace; Telson (tail); Abdomen) (exception to group: has 3 body parts)

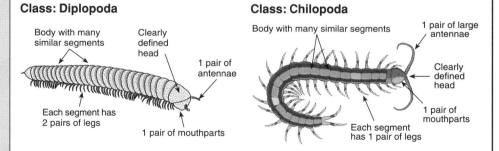

Class: Insecta — Honey bee (1 pair of antennae; 1 pair of compound eyes; 2 pairs of wings; 3 pairs of legs; Head; Thorax; Abdomen); Locust; Butterfly; Beetle. Beetles are the largest group within the animal kingdom with more than 300 000 species.

Class: Diplopoda — Body with many similar segments; Clearly defined head; 1 pair of antennae; Each segment has 2 pairs of legs; 1 pair of mouthparts

Class: Chilopoda — Body with many similar segments; 1 pair of large antennae; Clearly defined head; 1 pair of mouthparts; Each segment has 1 pair of legs

Phylum: Echinodermata

- Rigid body wall, internal skeleton made of calcareous plates.
- Many possess spines.
- Ventral mouth, dorsal anus.
- External fertilization.
- Unsegmented, marine organisms.
- Tube feet for locomotion.
- Water vascular system.

Examples: Starfish, brittlestars, feather stars, sea urchins, sea lilies.
Species diversity: 6000 +

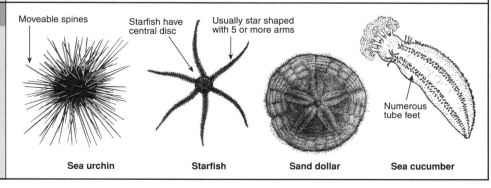

Sea urchin (Moveable spines); Starfish (Starfish have central disc; Usually star shaped with 5 or more arms); Sand dollar; Sea cucumber (Numerous tube feet)

Classification of Organisms

© Biozone International 2006-2007
Photocopying Prohibited

Kingdom: ANIMALIA (continued)

Phylum: Chordata

- Dorsal notochord (flexible, supporting rod) present at some stage in the life history.
- Post-anal tail present at some stage in their development.
- Dorsal, tubular nerve cord.
- Pharyngeal slits present.
- Circulation system closed in most.
- Heart positioned on ventral side.

Species diversity: 48 000 +

- A very diverse group with several sub-phyla:
  - Urochordata (sea squirts, salps)
  - Cephalochordata (lancelet)
  - Craniata (vertebrates)

Sub-Phylum Craniata (vertebrates)
- Internal skeleton of cartilage or bone.
- Well developed nervous system.
- Vertebral column replaces notochord.
- Two pairs of appendages (fins or limbs) attached to girdles.

Further subdivided into:

Class: Chondrichthyes (cartilaginous fish)
- Skeleton of cartilage (not bone).
- No swim bladder.
- All aquatic (mostly marine).
- Include: Sharks, rays, and skates.

Species diversity: 850 +

Class: Osteichthyes (bony fish)
- Swim bladder present.
- All aquatic (marine and fresh water).

Species diversity: 21 000 +

Class: Amphibia (amphibians)
- Lungs in adult, juveniles may have gills (retained in some adults).
- Gas exchange also through skin.
- Aquatic and terrestrial (limited to damp environments).
- Include: Frogs, toads, salamanders, and newts.

Species diversity: 3900 +

Class Reptilia (reptiles)
- Ectotherms with no larval stages.
- Teeth are all the same type.
- Eggs with soft leathery shell.
- Mostly terrestrial.
- Include: Snakes, lizards, crocodiles, turtles, and tortoises.

Species diversity: 7000 +

Class: Aves (birds)
- Terrestrial endotherms.
- Eggs with hard, calcareous shell.
- Strong, light skeleton.
- High metabolic rate.
- Gas exchange assisted by air sacs.

Species diversity: 8600 +

Class: Mammalia (mammals)
- Endotherms with hair or fur.
- Mammary glands produce milk.
- Glandular skin with hair or fur.
- External ear present.
- Teeth are of different types.
- Diaphragm between thorax/abdomen.

Species diversity: 4500 +

Subdivided into three subclasses: *Monotremes, marsupials, placentals.*

Class: Chondrichthyes (cartilaginous fish)

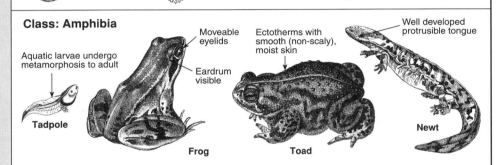

Hammerhead shark — Ectotherms with endoskeleton made of cartilage; Lateral line sense organ; Asymmetrical tail fin provides lift; Skin with toothlike scales; Pelvic fin; Pectoral fin; No operculum (bony flap) over gills. Stingray.

Class: Osteichthyes (bony fish)

Eel; Seahorse; Herring — Fins supported by bony rays; Slippery skin with thin, bony scales; Tail fin is symmetrical in shape; Operculum (bony flap) over gills; Sensory lateral line system; Ectotherms with bony endoskeleton.

Class: Amphibia

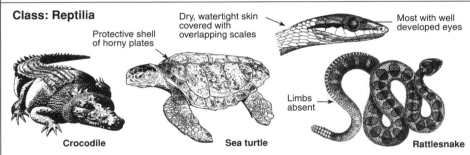

Tadpole — Aquatic larvae undergo metamorphosis to adult. Frog — Moveable eyelids; Eardrum visible. Toad — Ectotherms with smooth (non-scaly), moist skin. Newt — Well developed protrusible tongue.

Class: Reptilia

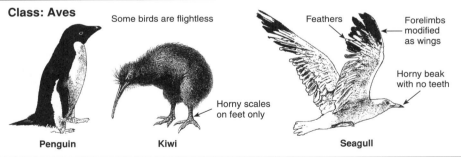

Crocodile. Sea turtle — Protective shell of horny plates; Dry, watertight skin covered with overlapping scales. Rattlesnake — Most with well developed eyes; Limbs absent.

Class: Aves

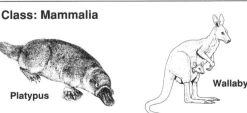

Penguin — Some birds are flightless. Kiwi — Horny scales on feet only. Seagull — Feathers; Forelimbs modified as wings; Horny beak with no teeth.

Class: Mammalia

Platypus — **Monotremes**: Egg laying mammals. Wallaby — **Marsupials**: Give birth to live, very immature young which then develop in a pouch. Wildebeest, Dolphin — **Placentals**: Have a placenta and give birth to live, well developed young.

© Biozone International 2006-2007

Classification System

The classification of organisms is designed to reflect how they are related to each other. The fundamental unit of classification of living things is the **species**. Its members are so alike genetically that they can interbreed. This genetic similarity also means that they are almost identical in their physical and other characteristics. Species are classified further into larger, more comprehensive categories (higher taxa). It must be emphasized that all such higher classifications are human inventions to suit a particular purpose.

1. The table below shows part of the classification for humans using the seven major levels of classification. For this question, use the example of the classification of the Ethiopean hedgehog, on the following page, as a guide.

   (a) Complete the list of the classification levels on the left hand side of the table below:

	Classification level	Human classification
1.	_____	_____
2.	_____	_____
3.	_____	_____
4.	_____	_____
5.	Family	Hominidae
6.	_____	_____
7.	_____	_____

   (b) The name of the Family that humans belong to has already been entered into the space provided. Complete the classification for humans (*Homo sapiens*) on the table above.

2. Describe the two-part scientific naming system (called the **binomial system**) that is used to name organisms:
   _____
   _____

3. Give two reasons explaining why the classification of organisms is important:

   (a) _____
   _____

   (b) _____
   _____

4. Traditionally, the classification of organisms has been based largely on similarities in physical appearance. More recently, new methods involving biochemical comparisons have been used to provide new insights into how species are related. Describe an example of a biochemical method for comparing how species are related:
   _____
   _____
   _____

5. As an example of physical features being used to classify organisms, mammals have been divided into three major subclasses: monotremes, marsupials, and placentals. Describe the main physical feature distinguishing each of these taxa:

   (a) Monotreme: _____
   _____

   (b) Marsupial: _____
   _____

   (c) Placental: _____
   _____

Related activities: New Classification Schemes, Classification Keys, Features of Taxonomic Groups

Classification of the Ethiopian Hedgehog

Below is the classification for the **Ethiopian hedgehog**. Only one of each group is subdivided in this chart showing the levels that can be used in classifying an organism. Not all possible subdivisions have been shown here. For example, it is possible to indicate such categories as **super-class** and **sub-family**. The only natural category is the **species**, often separated into geographical **races**, or **sub-species**, which generally differ in appearance.

Kingdom: **Animalia**
Animals; one of five kingdoms

Phylum: **Chordata**
Animals with a notochord (supporting rod of cells along the upper surface)
tunicates, salps, lancelets, and vertebrates

23 other phyla

Sub-phylum: **Vertebrata**
Animals with backbones
fish, amphibians, reptiles, birds, mammals

Class: **Mammalia**
Animals that suckle their young on milk from mammary glands
placentals, marsupials, monotremes

Sub-class: **Eutheria or Placentals**
Mammals whose young develop for some time in the female's reproductive tract gaining nourishment from a placenta
placental mammals

Order: **Insectivora**
Insect eating mammals
An order of over 300 species of primitive, small mammals that feed mainly on insects and other small invertebrates.

17 other orders

Sub-order: **Erinaceomorpha**
The hedgehog-type insectivores. One of the three suborders of insectivores. The other suborders include the tenrec-like insectivores (*tenrecs and golden moles*) and the shrew-like insectivores (*shrews, moles, desmans, and solenodons*).

Family: **Erinaceidae**
The only family within this suborder. Comprises two subfamilies: the true or spiny hedgehogs and the moonrats (gymnures). Representatives in the family include the common European hedgehog, desert hedgehog, and the moonrats.

Genus: ***Paraechinus***
One of eight genera in this family. The genus *Paraechinus* includes three species which are distinguishable by a wide and prominent naked area on the scalp.

7 other genera

Species: ***aethiopicus***
The Ethiopian hedgehog inhabits arid coastal areas. Their diet consists mainly of insects, but includes small vertebrates and the eggs of ground nesting birds.

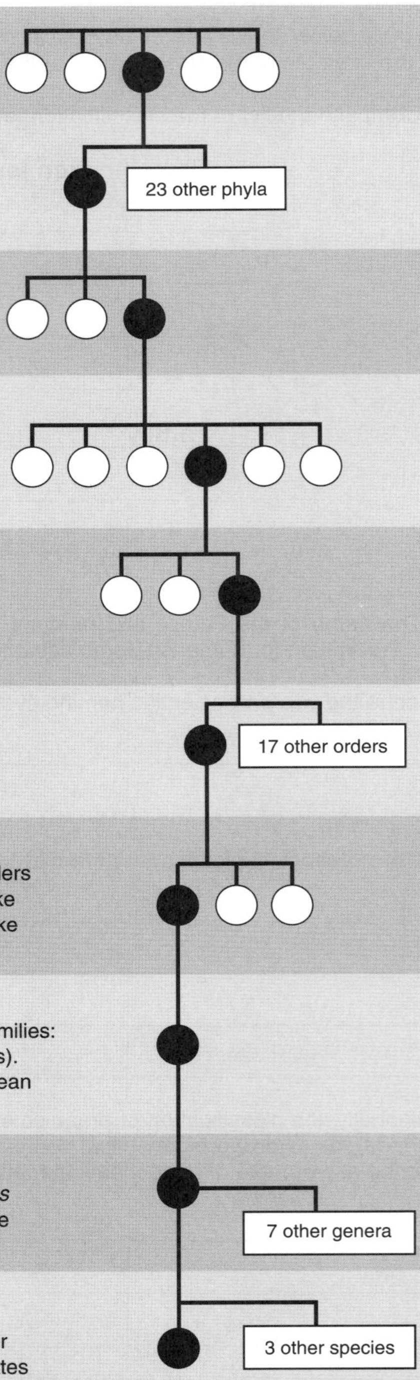

3 other species

The order *Insectivora* was first introduced to group together shrews, moles, and hedgehogs. It was later extended to include tenrecs, golden moles, desmans, tree shrews, and elephant shrews and the taxonomy of the group became very confused. Recent reclassification of the elephant shrews and tree shrews into their own separate orders has made the Insectivora a more cohesive group taxonomically.

Ethiopian hedgehog
Paraechinus aethiopicus

Classification Keys

Classification systems provide biologists with a way in which to identify species. They also indicate how closely related, in an evolutionary sense, each species is to others. An organism's classification should include a clear, unambiguous **description**, an accurate **diagram**, and its unique name, denoted by the **genus** and **species**. Classification keys are used to identify an organism and assign it to the correct species (assuming that the organism has already been formally classified and is included in the key). Typically, keys are **dichotomous** and involve a series of linked steps. At each step, a choice is made between two features; each alternative leads to another question until an identification is made. If the organism cannot be identified, it may be a new species or the key may need revision. Two examples of **dichotomous keys** are provided here. The first (below) describes features for identifying the larvae of various genera within the order Trichoptera (caddisflies). From this key you should be able to assign a generic name to each of the caddisfly larvae pictured. The key on the next page identifies aquatic insect orders.

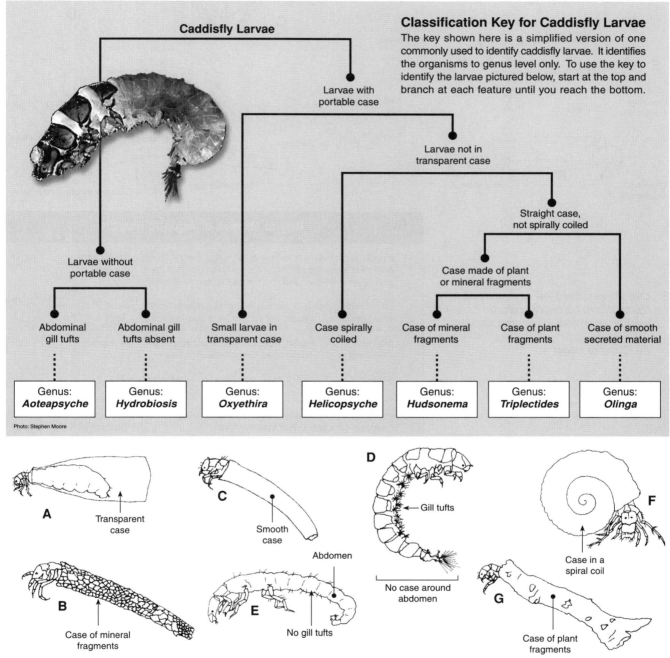

Classification Key for Caddisfly Larvae
The key shown here is a simplified version of one commonly used to identify caddisfly larvae. It identifies the organisms to genus level only. To use the key to identify the larvae pictured below, start at the top and branch at each feature until you reach the bottom.

1. Describe the main feature used to distinguish the genera in the key above: _____
_____

2. Use the key above to assign each of the caddisfly larvae (**A-G**) to its correct genus:

   A: _____   D: _____   G: _____

   B: _____   E: _____

   C: _____   F: _____

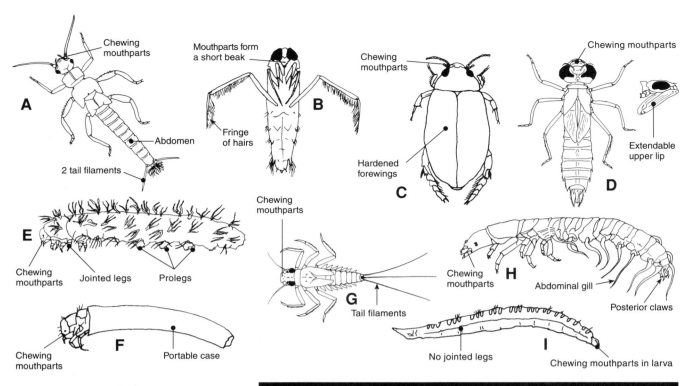

Key to Orders of Aquatic Insects

1	Insects with chewing mouthparts; forewings are hardened and meet along the midline of the body when at rest (they may cover the entire abdomen or be reduced in length).	**Coleoptera** (beetles)
	Mouthparts piercing or sucking and form a pointed cone	Go to 2
	With chewing mouthparts, but without hardened forewings	Go to 3
2	Mouthparts form a short, pointed beak; legs fringed for swimming or long and spaced for suspension on water.	**Hemiptera** (bugs)
	Mouthparts do not form a beak; legs (if present) not fringed or long, or spaced apart.	Go to 3
3	Prominent upper lip (labium) extendable, forming a food capturing structure longer than the head.	**Odonata** (dragonflies & damselflies)
	Without a prominent, extendable labium	Go to 4
4	Abdomen terminating in three tail filaments which may be long and thin, or with fringes of hairs.	**Ephemeroptera** (mayflies)
	Without three tail filaments	Go to 5
5	Abdomen terminating in two tail filaments	**Plecoptera** (stoneflies)
	Without long tail filaments	Go to 6
6	With three pairs of jointed legs on thorax	Go to 7
	Without jointed, thoracic legs (although non-segmented prolegs or false legs may be present).	**Diptera** (true flies)
7	Abdomen with pairs of non-segmented prolegs bearing rows of fine hooks.	**Lepidoptera** (moths and butterflies)
	Without pairs of abdominal prolegs	Go to 8
8	With eight pairs of finger-like abdominal gills; abdomen with two pairs of posterior claws.	**Megaloptera** (dobsonflies)
	Either, without paired, abdominal gills, or, if such gills are present, without posterior claws.	Go to 9
9	Abdomen with a pair of posterior prolegs bearing claws with subsidiary hooks; sometimes a portable case.	**Trichoptera** (caddisflies)

3. Use the simplified key to identify each of the orders (by order or common name) of aquatic insects (**A-I**) pictured above:

(a) Order of insect A: _____

(b) Order of insect B: _____

(c) Order of insect C: _____

(d) Order of insect D: _____

(e) Order of insect E: _____

(f) Order of insect F: _____

(g) Order of insect G: _____

(h) Order of insect H: _____

(i) Order of insect I: _____

Keying Out Plant Species

Dichotomous keys are a useful tool in biology and can enable identification to the species level provided the characteristics chosen are appropriate for separating species. Keys are extensively used by botanists as they are quick and easy to use in the field, although they sometimes rely on the presence of particular plant parts such as fruits or flowers. Some also require some specialist knowledge of plant biology. The following simple activity requires you to identify five species of the genus *Acer* from illustrations of the leaves. It provides valuable practice in using characteristic features to identify plants to species level.

A Dichotomous Key to Some Common Maple Species

1a Adult leaves with five lobes 2
1b Adult leaves with three lobes 4
    2a Leaves 7.5-13 cm wide, with smooth edges, lacking serrations along the margin. U shaped sinuses between lobes.
        Sugar maple, *Acer saccharum*
    2b Leaves with serrations (fine teeth) along the margin 3
        3a Leaves 5-13 cm wide and deeply lobed.
            Japansese maple, *Acer palmatum*
        3b Leaves 13-18 cm wide and deeply lobed.
            Silver maple, *Acer saccharinum*
        4a Leaves 5-15 cm wide with small sharp serrations on the margins. Distinctive V shaped sinuses between the lobes.
            Red maple, *Acer rubrum*
        4b Leaves 7.5-13 cm wide without serrations on the margins. Shallow sinuses between the lobes.
            Black maple, *Acer nigrum*

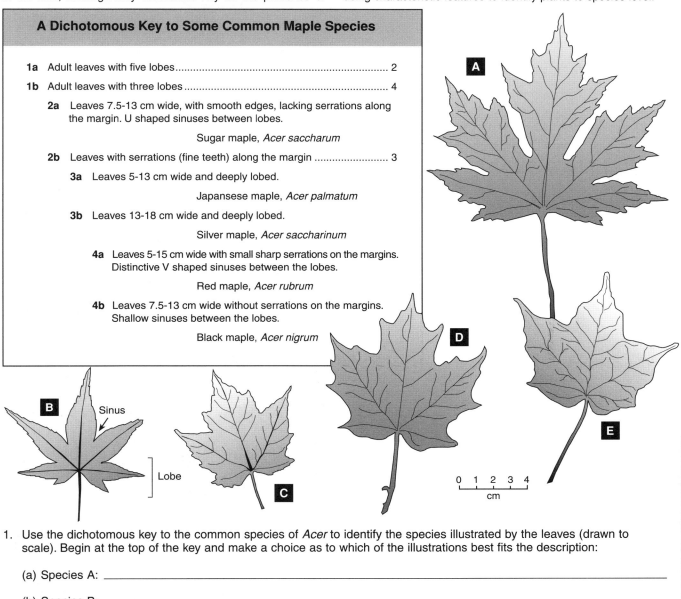

1. Use the dichotomous key to the common species of *Acer* to identify the species illustrated by the leaves (drawn to scale). Begin at the top of the key and make a choice as to which of the illustrations best fits the description:

   (a) Species A: _____

   (b) Species B: _____

   (c) Species C: _____

   (d) Species D: _____

   (e) Species E: _____

2. Identify a feature that could be used to identify maple species when leaves are absent: _____

3. Suggest why it is usually necessary to consider a number of different features in order to classify plants to species level:
   _____
   _____

4. When identifying a plant, suggest what you should be sure of before using a key to classify it to species level:
   _____
   _____

Features of the Five Kingdoms

The classification of living things into taxonomic groups is based on how biologists believe they are related in an evolutionary sense. Organisms in a taxonomic group share common features that set the group apart from others. By identifying these **distinguishing features**, it is possible to gain an understanding of the evolutionary development of the group. The focus of this activity is to summarize the distinguishing features of each of the five kingdoms.

1. Distinguishing features of Kingdom **Prokaryotae**:

   _____
   _____
   _____
   _____
   _____

2. Distinguishing features of Kingdom **Protista**:

   _____
   _____
   _____
   _____
   _____

3. Distinguishing features of Kingdom **Fungi**:

   _____
   _____
   _____
   _____
   _____

4. Distinguishing features of Kingdom **Plantae**:

   _____
   _____
   _____
   _____
   _____

5. Distinguishing features of Kingdom **Animalia**:

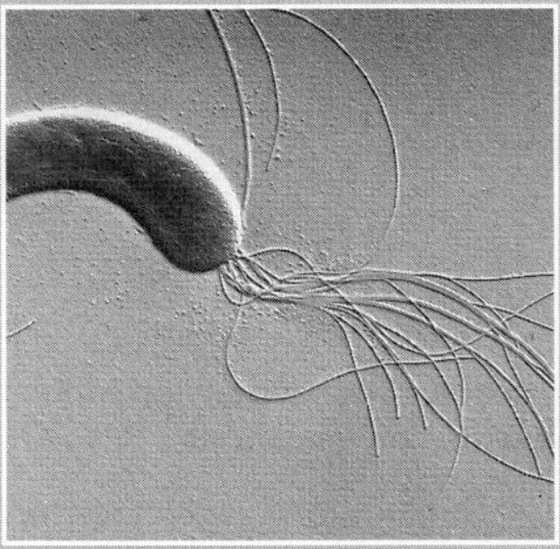

   _____
   _____
   _____
   _____
   _____

Spirillum bacteria 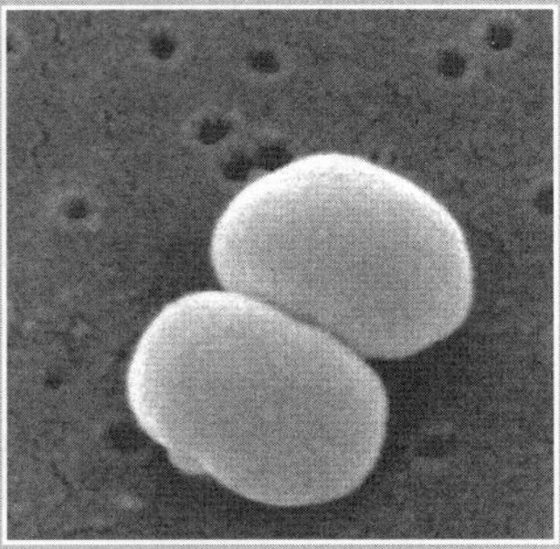 | *Staphylococcus*

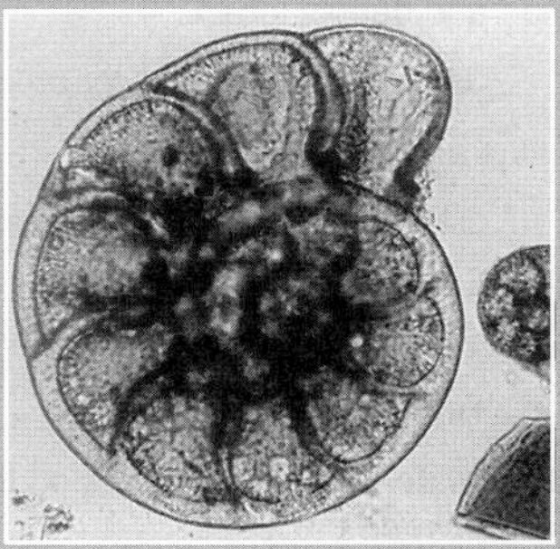

Foraminiferan 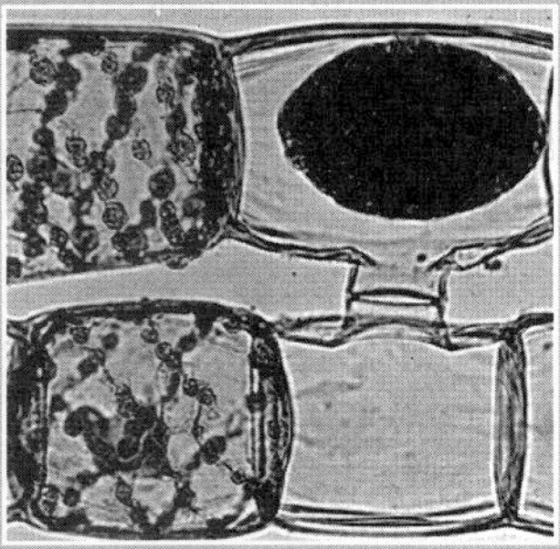 | *Spirogyra* algae

Mushrooms 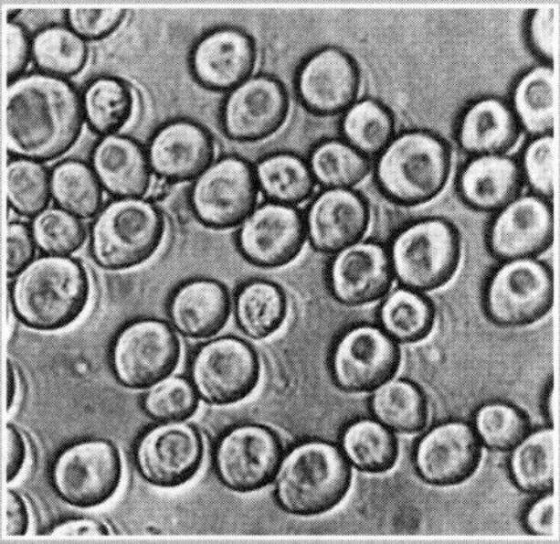 | Yeast cells in solution

Moss  | Pea plants

Cicada moulting  | Gibbon

RA 1 **Related activities**: The New Tree of Life, Features of Taxonomic Groups

© Biozone International 2006-2007
Photocopying Prohibited

The Classification of Life

For this activity, cut away the two pages of diagrams that follow from your book. The five kingdoms that all living things are grouped into, are listed on this page and the following page.

1. Cut out all of the images of different living organisms (cut around each shape closely, taking care to include their names).
2. Sort them into their classification groups by placing them into the spaces provided on this and the following page.
3. To fix the images in place, first use a temporary method (e.g. a gluestick or sellotape folded into a loop), so that you can easily reposition them if you need to. Make a permanent fixture when you are completely satisfied with your placements on the page.

Bacteria

Protists

Fungi

Plants			
Bryophytes	**Seedless plants**	**Angiosperms: Monocotyledon**	**Angiosperms: Dicotyledon**
	Gymnosperms: Cycads	**Gymnosperms: Conifers**	

Related activities: Features of Taxonomic Groups

Animals				
Sponges	**Cnidarians**		**Flatworms**	**Annelids**

Molluscs: Gastropods   Bivalves   Cephalopods   **Echinoderms**

Arthropods: Crustaceans   Myriapods   Arachnids   Insects

Cartilaginous fish	**Bony fish**	**Amphibians**

Reptiles	**Birds**

Mammals: Monotremes   Marsupials   Placentals

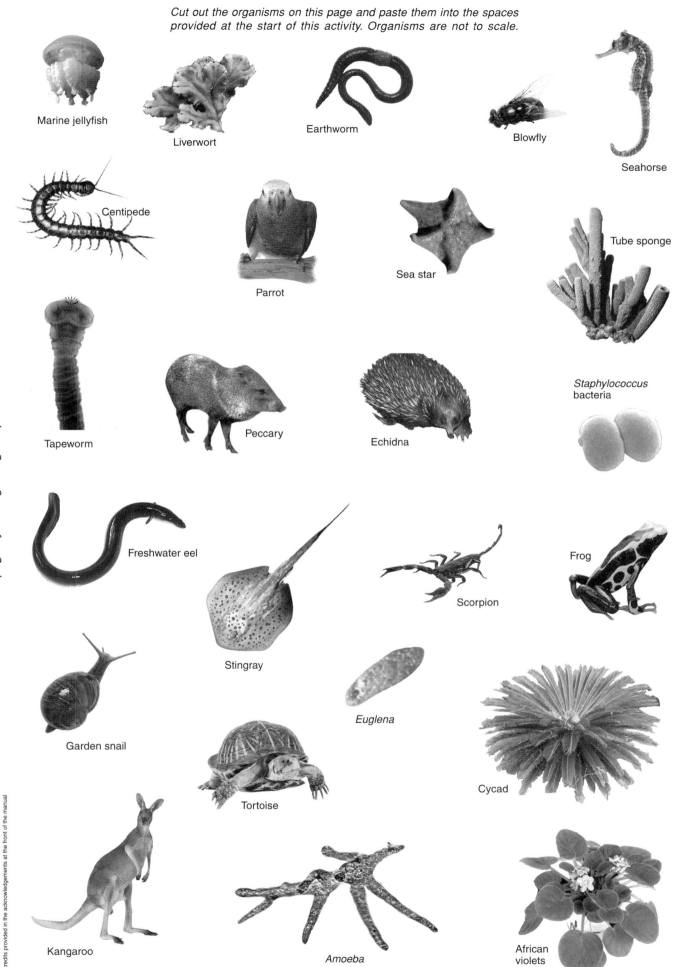

Cut out the organisms on this page and paste them into the spaces provided at the start of this activity. Organisms are not to scale.

This page has been deliberately left blank

Cut out the images
on the other side of this page

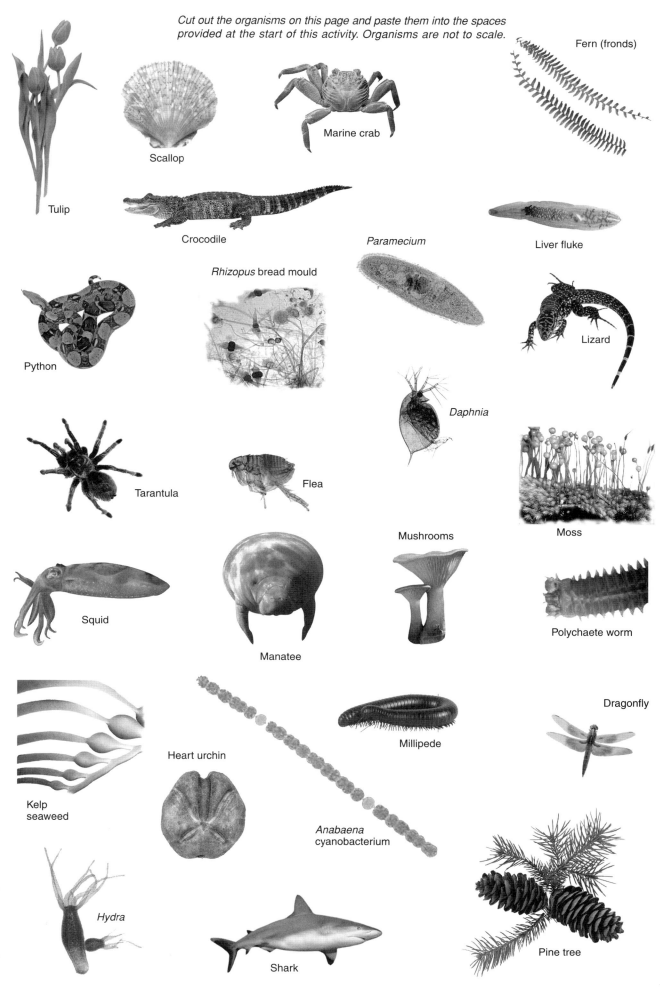

This page has been deliberately left blank

Cut out the images
on the other side of this page

Features of Animal Taxa

The animal kingdom is classified into about 35 major **phyla**. Representatives of the more familiar taxa are illustrated below: **cnidarians** (includes jellyfish, sea anemones, and corals), **annelids** (segmented worms), **arthropods** (insects, crustaceans, spiders, scorpions, centipedes and millipedes), **molluscs** (snails, bivalve shellfish, squid and octopus), **echinoderms** (starfish and sea urchins), **vertebrates** from the phylum **chordates** (fish, amphibians, reptiles, birds, and mammals). The **arthropods** and the **vertebrates** have been represented in more detail, giving the **classes** for each of these **phyla**. This activity asks you to describe the **distinguishing features** of each of the taxa represented below.

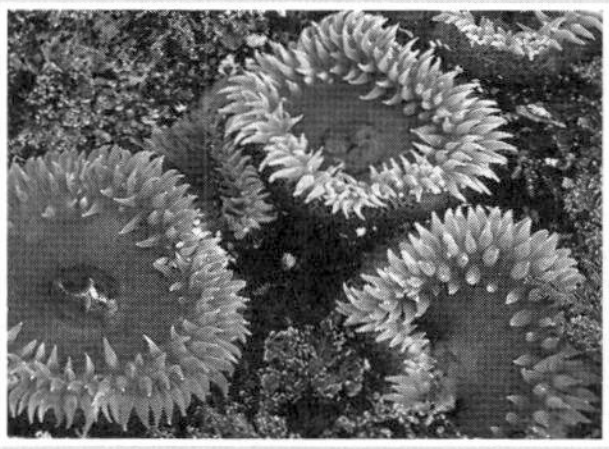
Sea anemones

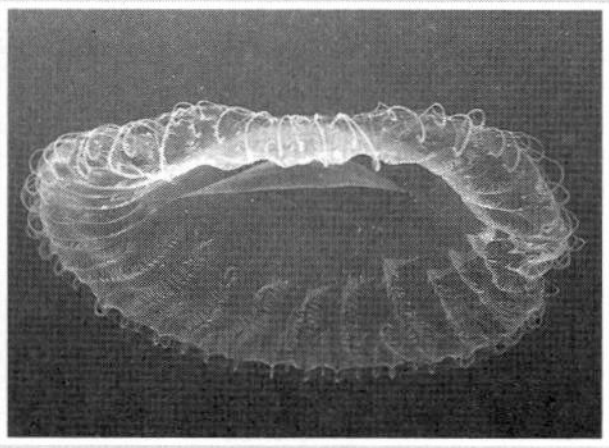
Jellyfish

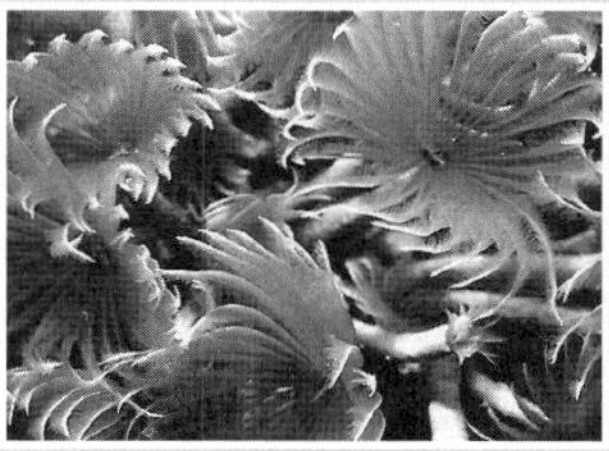
Tubeworms

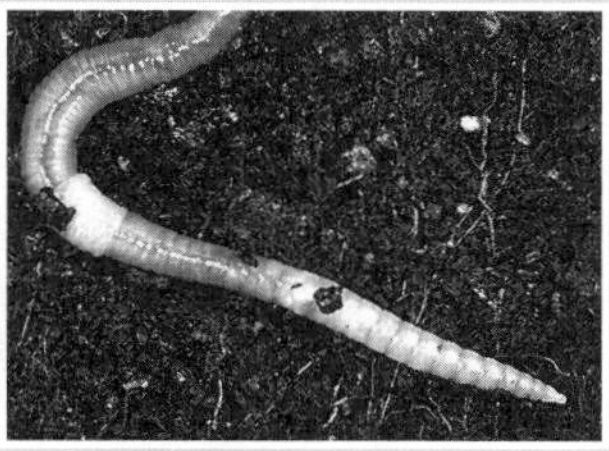
BOB

Earthworm

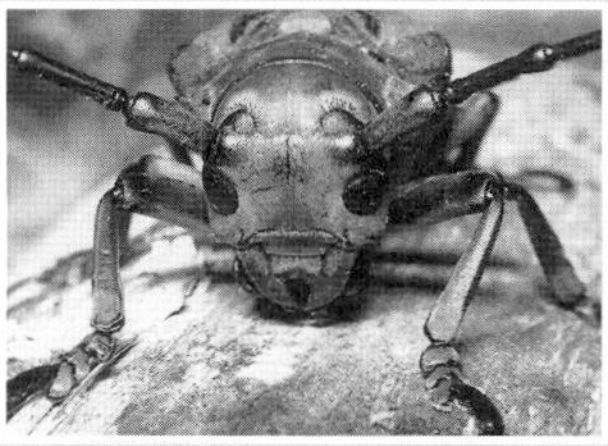
Long-horned beetle

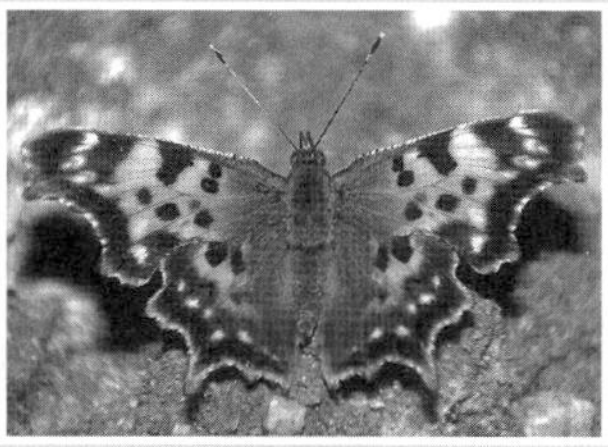
Butterfly

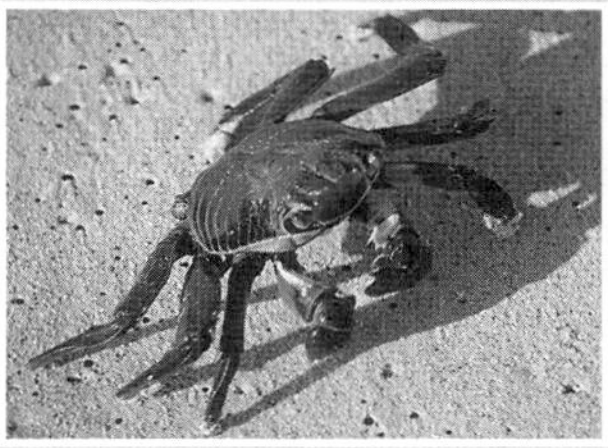
Crab

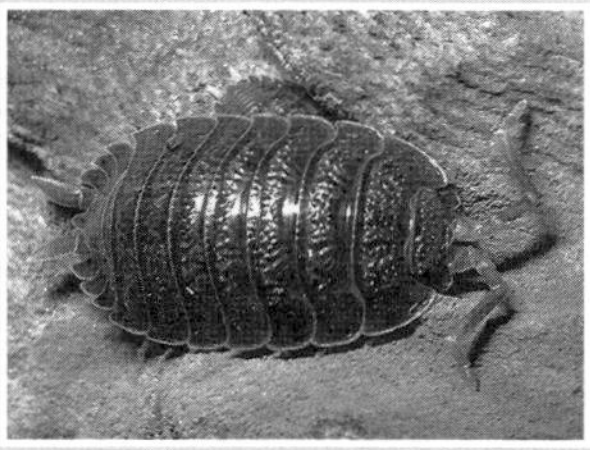
Woodlouse

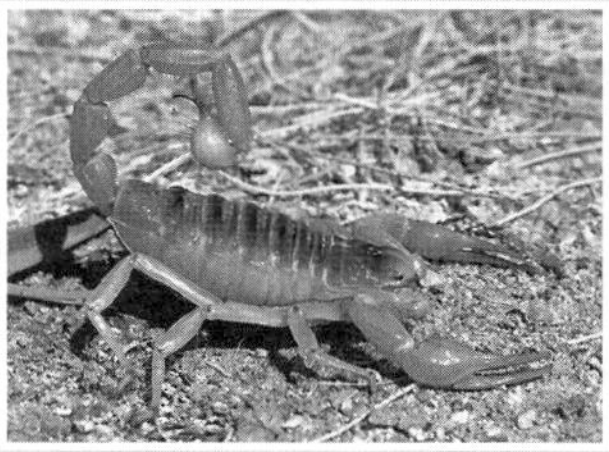
Scorpion

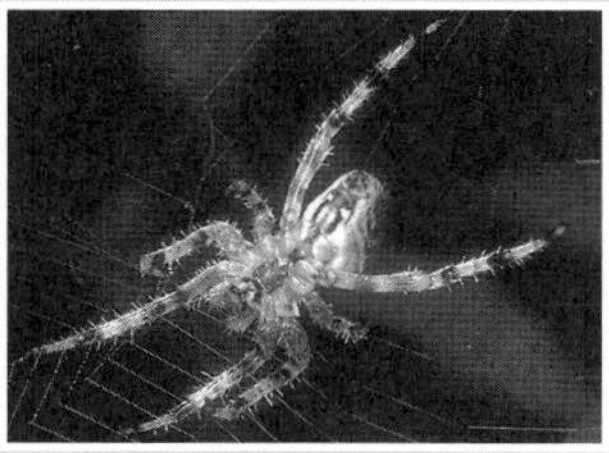
Spider

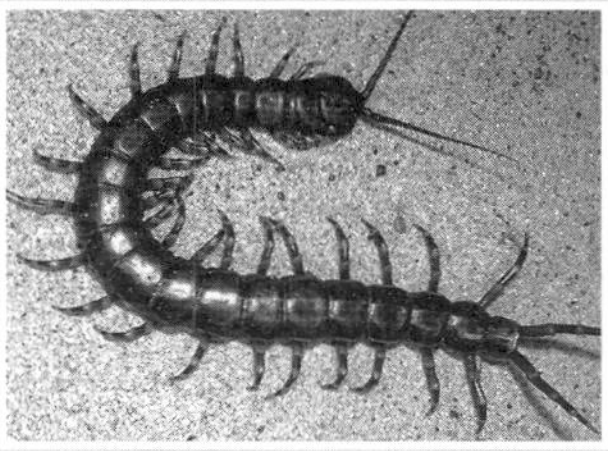
Centipede

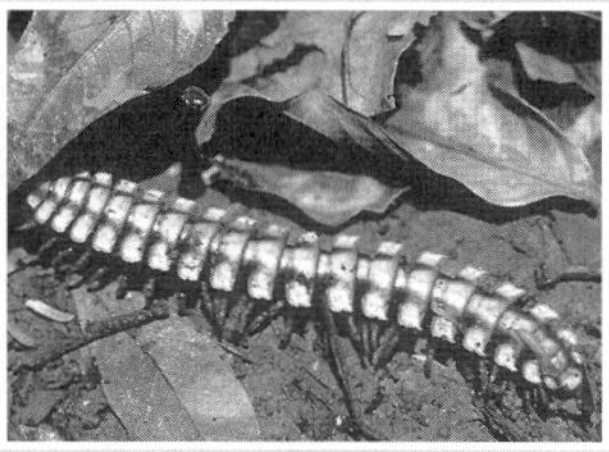
Millipede

1. **Cnidarian** features:

2. **Annelid** features:

3. **Insect** features:

4. **Crustacean** features:

5. **Arachnid** features:

6. **Myriapod** (class Chilopoda and Diplopoda) features:

Related activities: Features of Taxonomic Groups

7. **Mollusc** features: _____

8. **Echinoderm** features: _____

9. **Fish** features: _____

10. **Amphibian** features: _____

11. **Reptile** features: _____

12. **Bird** features: _____

13. **Mammal** features: _____

Features of Macrofungi and Plants

Although plants and fungi are some of the most familiar organisms in our environment, their classification has not always been straightforward. We know now that the plant kingdom is monophyletic, meaning that it is derived from a common ancestor. The variety we see in plant taxa today is a result of their enormous diversification from the first plants. Although the fungi were once grouped together with the plants, they are unique organisms that differ from other eukaryotes in their mode of nutrition, structural organization, growth, and reproduction. The focus of this activity is to summarize the features of the fungal kingdom, the major divisions of the plant kingdom, and the two classes of flowering plants (angiosperms).

1. **Macrofungi** features:
   _____
   _____
   _____
   _____

2. **Moss and liverwort** features:
   _____
   _____
   _____
   _____

3. **Fern** features:
   _____
   _____
   _____
   _____

4. **Gymnosperm** features:
   _____
   _____
   _____
   _____

5. **Monocot angiosperm** features:
   _____
   _____
   _____
   _____

6. **Dicot angiosperm** features:
   _____
   _____
   _____
   _____

Laboratory Techniques

Developing skills in the use of laboratory equipment, including safety in microbial techniques

Microscopy, biological drawings, differential centrifugation, biochemical tests, gel electrophoresis, tissue culture, microbiological techniques.

Learning Objectives

☐ 1. Compile your own glossary from the **KEY WORDS** displayed in **bold type** in the learning objectives below.

Microscopy *(pages 107-115)*

☐ 2. Distinguish between the structure and function of **optical** and **electron microscopes**. With reference to light and electron microscopy, explain and distinguish between: **magnification** and **resolution**.

☐ 3. Distinguish between the basic structure and operation of **transmission electron microscopes** (TEM) and **scanning electron microscopes** (SEM).

☐ 4. Distinguish between **compound** and **stereo light microscopes** and identify situations in which each would be used. Demonstrate an ability to use both these microscopes to locate material and focus images.

☐ 5. List the steps required for preparing a **temporary mount** for viewing with a compound light microscope. If required, demonstrate an ability to use **oil immersion** to view material in microscopy. State when it should (and should not) be used.

☐ 6. Demonstrate an ability to use simple **staining techniques** to show specific features of cells. Explain why **stains** are useful in the preparation of specimens.

☐ 7. Describe and interpret drawings and photographs of typical plant and animal cells (e.g. leaf palisade cell and liver cell) as seen using electron microscopy.

☐ 8. Make good **biological drawings** as a way of recording information where appropriate to your investigation.

Basic Lab Procedures *(pages 116-119, 123)*

☐ 9. Explain the basis of **chromatography** as a technique for separating and identifying biological molecules. Describe the calculation and use of R_f **values**.

☐ 10. Describe the following tests: **Benedict's test** for distinguishing between **reducing** and **non-reducing sugars**, I_2/**KI** (*iodine in potassium iodide solution*) **test** for starch, **emulsion test** for lipids, **biuret test** for proteins. Explain the basis of each test and its result.

☐ 11. Describe the principles of **differential centrifugation** (cell fractionation). Explain how it is achieved through **homogenization** followed by ultracentrifugation.

☐ 12. Explain the basic principles and role of **gel electrophoresis** (of DNA) in gene technology. Identify properties of the **gel** that facilitate the separation of DNA fragments.

☐ 13. Describe the techniques involved in **plant tissue culture**. Recognize tissue culture as a basic technique and discuss its applications in agriculture and forestry.

Microbial Techniques *(pages 120-122)*

☐ 14. Demonstrate an ability to use aseptic technique to prepare and inoculate **nutrient broths** (**liquid media**) and **nutrient agar** plates (**solid media**).

☐ 15. Demonstrate an ability to use **aseptic technique** to isolate a strain of bacteria using **streak plating**.

☐ 16. Compare methods used to measure growth in bacterial populations: **turbidimetry**, **dilution plating** (serial dilution), and **hemocytometer** counts. Distinguish between **total cell count** and **viable cell count** and discuss how the latter might be assessed.

See page 7 for additional details of this text:
■ Adds, J. *et al.*, 1999. **Tools, Techniques and Assessment in Biology** (NelsonThornes).

See page 7 for details of publishers of periodicals:

STUDENT'S REFERENCE

■ **Scanning Electron Microscopy** Biol. Sci. Rev., 13(3) January 2001, pp. 6-9. *An excellent account of the techniques and applications of SEM. Includes details of specimen preparation and recent advancements in the technology.*

■ **X-Ray Microscopy** Biol. Sci. Rev., 14(2) Nov. 2001, pp. 38-40. *The technique and application of X-ray microscopy and its advantages over EM.*

■ **Transmission Electron Microscopy** Biol. Sci. Rev., 13(2) Nov. 2000, pp. 32-35. *An account of the techniques and applications of TEM. Includes a diagram comparing features of TEM and LM.*

■ **Light Microscopy** Biol. Sci. Rev., 13(1) Sept. 2000, pp. 36-38. *An excellent account of the basis and various techniques of light microscopy.*

■ **Size Does Matter** Biol. Sci. Rev., 17 (3) February 2005, pp. 10-13. *Measuring the size of organisms and calculating magnification and scale.*

■ **Musings on Measurements** Biol. Sci. Rev., 16(2) Nov. 2003, pp. 30-32. *Accuracy of measurement and its importance in science.*

■ **Micropipetting** American Biology Teacher, 66(4), April 2004, pp. 291-296. *A how-to-do-it account of the use and care of a micropipettor. This account also describes activities designed to help biology students develop their skills in using this important laboratory tool for molecular biology.*

■ **Fast Tissue Culture** Biol. Sci. Rev., 10(3) January 1998, pp. 2-6. *The techniques and applications of simple tissue culture in plants.*

See pages 4-5 for details of how to access **Bio Links** from our web site: **www.thebiozone.com** From Bio Links, access sites under the topics:

CELL BIOLOGY AND BIOCHEMISTRY:
> **Microscopy**: • Biological applications of electron and light microscopy • Microscopy UK • Scanning Electron Microscope ... *and others*

BIOTECHNOLOGY > **Biotechnology Techniques**: • Gel electrophoresis slide presentation > **Applications in Biotechnology** > **Cloning and Tissue Culture**: • Tissue culture in the classroom

Presentation MEDIA to support this topic:
GENES & INHERITANCE
• Gene Technology

Biological Drawings

Microscopes are a powerful tool for examining cells and cell structures. In order to make a permanent record of what is seen when examining a specimen, it is useful to make a drawing. It is important to draw **what is actually seen**. This will depend on the **resolution** of the microscope being used. Resolution refers to the ability of a microscope to separate small objects that are very close together. Making drawings from mounted specimens is a skill. Drawing forces you to observe closely and accurately. While photographs are limited to representing appearance at a single moment in time, drawings can be composites of the observer's cumulative experience, with many different specimens of the same material. The total picture of an object thus represented can often communicate information much more effectively than a photograph. Your attention to the outline of suggestions below will help you to make more effective drawings. If you are careful to follow the suggestions at the beginning, the techniques will soon become habitual.

1. **Drawing materials**: All drawings should be done with a clear pencil line on good quality paper. A sharp HB pencil is recommended. A soft rubber of good quality is essential. Diagrams in ballpoint or fountain pen are unacceptable because they cannot be corrected.

2. **Positioning**: Centre your diagram on the page. Do not draw it in a corner. This will leave plenty of room for the addition of labels once the diagram is completed.

3. **Size**: A drawing should be large enough to easily represent all the details you see without crowding. Rarely, if ever, are drawings too large, but they are often too small. Show only as much as is necessary for an understanding of the structure; a small section shown in detail will often suffice. It is time consuming and unnecessary, for example, to reproduce accurately the entire contents of a microscope field.

4. **Accuracy**: Your drawing should be a complete, accurate representation of the material you have observed, and should communicate your understanding of the material to anyone who looks at it. Avoid making "idealized" drawings; your drawing should be a picture of what you actually see, not what you imagine should be there. Proportions should be accurate. If necessary, measure the lengths of various parts with a ruler. If viewing through a microscope, estimate them as a proportion of the field of view, then translate these proportions onto the page. When drawing shapes that indicate an outline, make sure the line is complete. Where two ends of a line do not meet (as in drawing a cell outline) then this would indicate that it has a hole in it.

5. **Technique**: Use only simple, narrow lines. Represent depth by stippling (dots close together). Indicate depth only when it is essential to your drawing (usually it is not). Do not use shading. Look at the specimen while you are drawing it.

6. **Labels**: Leave a good margin for labels. All parts of your diagram must be labelled accurately. Labelling lines should be drawn with a ruler and should not cross. Where possible, keep label lines vertical or horizontal. Label the drawing with:
   - A title, which should identify the material (organism, tissues or cells).
   - Magnification under which it was observed, or a scale to indicate the size of the object.
   - Names of structures.
   - In living materials, any movements you have seen.

Remember that drawings are intended as records for you, and as a means of encouraging close observation; artistic ability is not necessary. Before you turn in a drawing, ask yourself if you know what every line represents. If you do not, look more closely at the material. *Take into account the rules for biological drawings and draw what you see, not what you think you see!*

Examples of acceptable biological drawings: The diagrams below show two examples of biological drawings that are acceptable. The example on the left is of a whole organism and its size is indicated by a scale. The example on the right is of plant tissue – a group of cells that are essentially identical in the structure. It is not necessary to show many cells even though your view through the microscope may show them. As few as 2-4 will suffice to show their structure and how they are arranged. Scale is indicated by stating how many times larger it has been drawn. Do not confuse this with what magnification it was viewed at under the microscope. The abbreviation **T.S.** indicates that the specimen was a cross or transverse section.

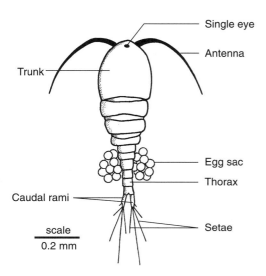

Cyclopoid copepod

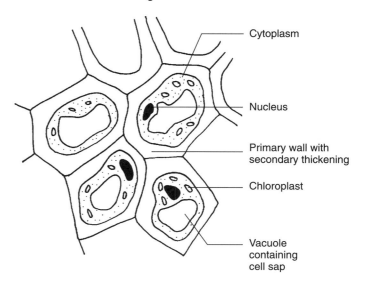

Collenchyma T.S. from Helianthus stem
Magnification x 450

An Unacceptable Biological Drawing

The diagram below is an example of how *not* to produce a biological drawing; it is based on the photograph to the left. There are many aspects of the drawing that are unacceptable. The exercise below asks you to identify the errors in this student's attempt.

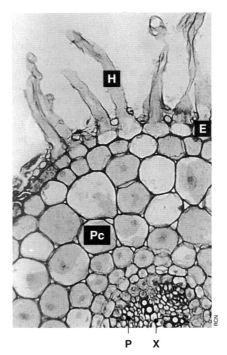

Specimen used for drawing

The photograph above is a light microscope view of a stained transverse section (cross section) of a root from a *Ranunculus* (buttercup) plant. It shows the arrangement of the different tissues in the root. The vascular bundle is at the center of the root, with the larger, central xylem vessels (**X**) and smaller phloem vessels (**P**) grouped around them. The root hair cells (**H**) are arranged on the external surface and form part of the epidermal layer (**E**). Parenchyma cells (**Pc**) make up the bulk of the root's mass. The distance from point **X** to point **E** on the photograph (above) is about 0.15 mm (150 μm).

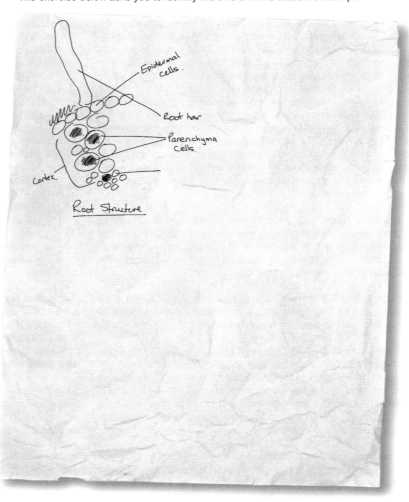

1. Identify and describe eight unacceptable features of the student's biological diagram above:

   (a) _____

   (b) _____

   (c) _____

   (d) _____

   (e) _____

   (f) _____

   (g) _____

   (h) _____

2. In the remaining space next to the 'poor example' (above) or on a blank piece of refill paper, attempt your own version of a biological drawing for the same material, based on the photograph above. Make a point of correcting all of the errors that you have identified in the sample student's attempt.

3. Explain why accurate biological drawings are more valuable to a scientific investigation than an 'artistic' approach:

   _____

   _____

   _____

Optical Microscopes

The light microscope is one of the most important instruments used in biology practicals, and its correct use is a basic and essential skill of biology. High power light microscopes use a combination of lenses to magnify objects up to several hundred times. They are called **compound microscopes** because there are two or more separate lenses involved. A typical compound light microscope (bright field) is shown below (top photograph). The specimens viewed with these microscopes must be thin and mostly transparent. Light is focused up through the condenser and specimen; if the specimen is thick or opaque, little or no detail will be visible. The microscope below has two eyepieces (**binocular**), although monocular microscopes, with a mirror rather than an internal light source, may still be encountered. Dissecting microscopes (lower photograph) are a type of binocular microscope used for observations at low total magnification (x4 to x50), where a large working distance between objectives and stage is required. A dissecting microscope has two separate lens systems, one for each eye. Such microscopes produce a 3-D view of the specimen and are sometimes called stereo microscopes for this reason.

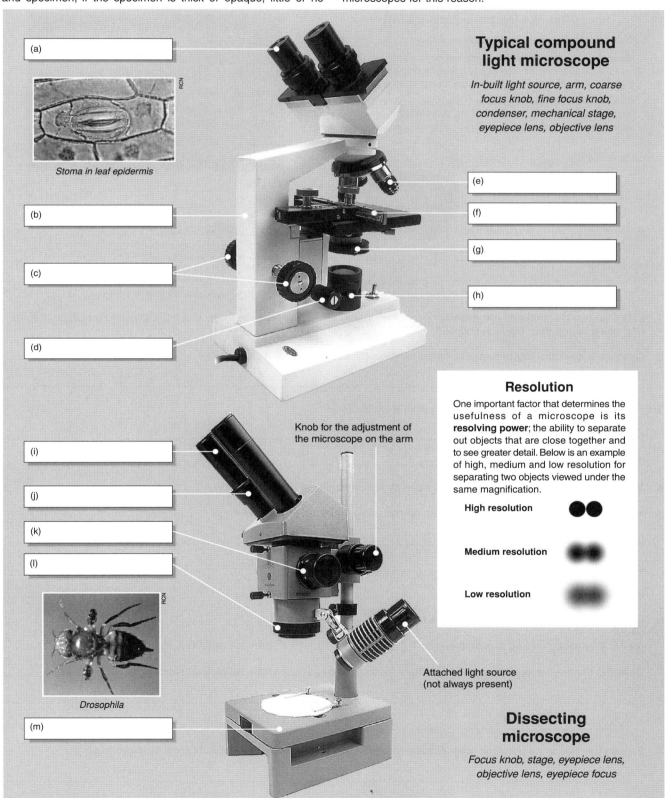

Typical compound light microscope

In-built light source, arm, coarse focus knob, fine focus knob, condenser, mechanical stage, eyepiece lens, objective lens

Stoma in leaf epidermis

Resolution

One important factor that determines the usefulness of a microscope is its **resolving power**; the ability to separate out objects that are close together and to see greater detail. Below is an example of high, medium and low resolution for separating two objects viewed under the same magnification.

High resolution

Medium resolution

Low resolution

Knob for the adjustment of the microscope on the arm

Attached light source (not always present)

Drosophila

Dissecting microscope

Focus knob, stage, eyepiece lens, objective lens, eyepiece focus

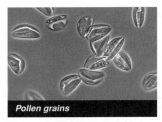
Pollen grains
Phase contrast illumination increases contrast of transparent specimens by producing interference effects.

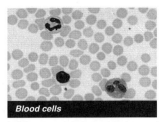
Blood cells
Leishman's stain is used to show red blood cells as red/pink, while staining the nucleus of white blood cells blue.

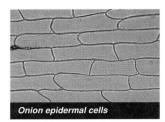
Onion epidermal cells
Standard **bright field** lighting shows cells with little detail; only cell walls, with the cell nuclei barely visible.

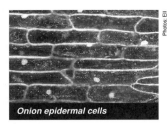
Onion epidermal cells
Dark field illumination is excellent for viewing near transparent specimens. The nucleus of each cell is visible.

Making a temporary wet mount

1. **Sectioning:** Very thin sections of fresh material are cut with a razorblade.
2. **Mounting:** The thin section(s) are placed in the center of a clean glass microscope slide and covered with a drop of mounting liquid (e.g. water, glycerol or stain). A coverslip is placed on top to exclude air (below).
3. **Staining:** Dyes can be applied to stain some structures and leave others unaffected. The stains used in dyeing living tissues are called **vital stains** and they can be applied before or after the specimen is mounted.

Commonly used temporary stains

Stain	Final color	Used for
Iodine solution	blue-black	Starch
Aniline sulfate	yellow	Lignin
Schultz's solution	blue	Starch
	blue or violet	Cellulose
	yellow	Protein, cutin, lignin, suberin
Methylene blue	blue	Nuclei

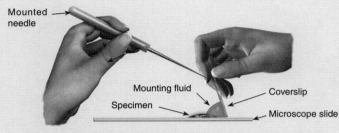
A mounted needle is used to support the coverslip and lower it gently over the specimen. This avoids including air in the mount.

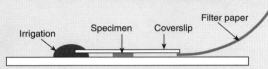
If a specimen is already mounted, a drop of stain can be placed at one end of the coverslip and drawn through using filter paper (above). Water can be drawn through in the same way to remove excess stain.

1. Label the two diagrams on the previous page, the bright field microscope (a) to (h) and the dissecting microscope (i) to (m), using words from the lists supplied.

2. Describe a situation where phase contrast microscopy would improve image quality: _____

3. List two structures that could be seen by light microscopy in:

   (a) A plant cell: _____

   (b) An animal cell: _____

4. Name one cell structure that can not be seen by light microscopy: _____

5. Identify a stain that would be appropriate for improving definition of the following:

   (a) Blood cells: _____   (d) Lignin: _____

   (b) Starch: _____   (e) Nuclei and DNA: _____

   (c) Protein: _____   (f) Cellulose: _____

6. Determine the magnification of a microscope using:

   (a) 15 X eyepiece and 40 X objective lens: _____   (b) 10 X eyepiece and 60 X objective lens: _____

7. Describe the main difference between a bright field light microscope and a dissecting microscope. _____

8. Explain the difference between magnification and resolution (resolving power) with respect to microscope use: _____

Electron Microscopes

Electron microscopes (EMs) use a beam of electrons, instead of light, to produce an image. The higher resolution of EMs is due to the shorter wavelengths of electrons. There are two basic types of electron microscope: **scanning electron microscopes** (SEM) and **transmission electron microscopes** (TEM). In SEMs, the electrons are bounced off the surface of an object to produce detailed images of the external appearance. TEMs produce very clear images of specially prepared thin sections.

Transmission Electron Microscope (TEM)

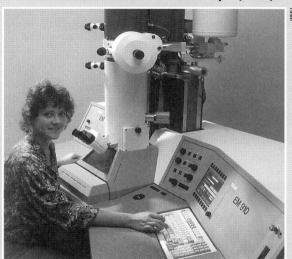

The transmission electron microscope is used to view extremely thin sections of material. Electrons pass through the specimen and are scattered. Magnetic lenses focus the image onto a fluorescent screen or photographic plate. The sections are so thin that they have to be prepared with a special machine, called an **ultramicrotome**, that can cut wafers to just 30 thousandths of a millimeter thick. It can magnify several hundred thousand times.

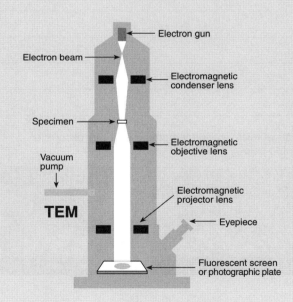

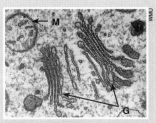

TEM photo showing the Golgi (**G**) and a mitochondrion (**M**).

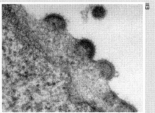

Three HIV viruses budding out of a human lymphocyte (TEM).

Scanning Electron Microscope (SEM)

The scanning electron microscope scans a sample with a beam of primary electrons that knock electrons from its surface. These secondary electrons are picked up by a collector, amplified, and transmitted onto a viewing screen or photographic plate, producing a superb 3-D image. A microscope of this power can easily obtain clear pictures of organisms as small as bacteria and viruses. The image produced is of the outside surface only.

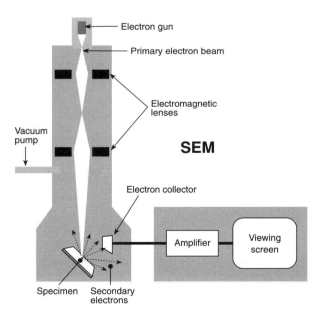

SEM photo of stoma and epidermal cells on the upper surface of a leaf.

Image of hair louse clinging to two hairs on a Hooker's sealion (SEM).

© Biozone International 2006-2007
Photocopying Prohibited

Related activities: Interpreting Electron Micrographs

	Light Microscope	Transmission Electron Microscope (TEM)	Scanning Electron Microscope (SEM)
Radiation source:	light	electrons	electrons
Wavelength:	400-700 nm	0.005 nm	0.005 nm
Lenses:	glass	electromagnetic	electromagnetic
Specimen:	living or non-living supported on glass slide	non-living supported on a small copper grid in a vacuum	non-living supported on a metal disc in a vacuum
Maximum resolution:	200 nm	1 nm	10 nm
Maximum magnification:	1500 x	250 000 x	100 000 x
Stains:	colored dyes	impregnated with heavy metals	coated with carbon or gold
Type of image:	colored	monochrome (black & white)	monochrome (black & white)

1. Explain why electron microscopes are able to resolve much greater detail than a light microscope:

2. Describe two typical applications for each of the following types of microscope:

    (a) Transmission electron microscope (TEM):

    (b) Scanning electron microscope (SEM):

    (c) Compound light microscope (thin section):

    (d) Dissecting microscope:

3. Identify which type of electron microscope (SEM or TEM) or optical microscope (compound light microscope or dissecting) was used to produce each of the images in the photos below (A-H):

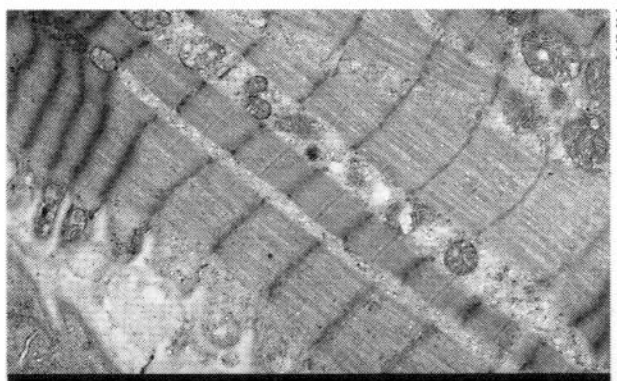

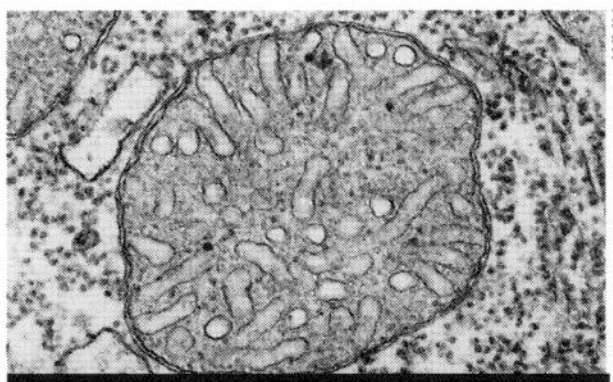

Cardiac muscle — A _____

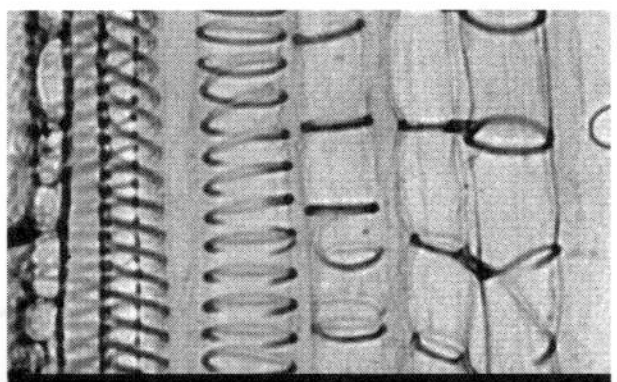

Plant vascular tissue — B _____

Mitochondrion — C _____

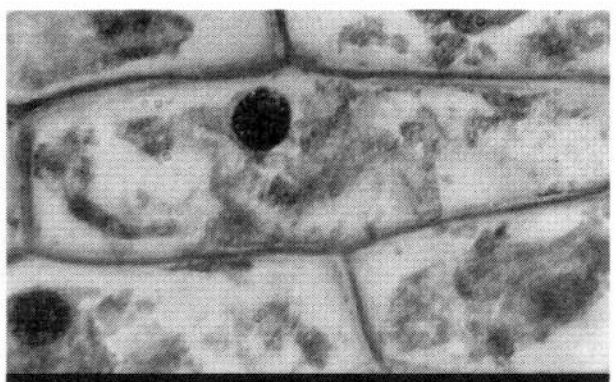

Plant epidermal cells — D _____

Head louse — E _____

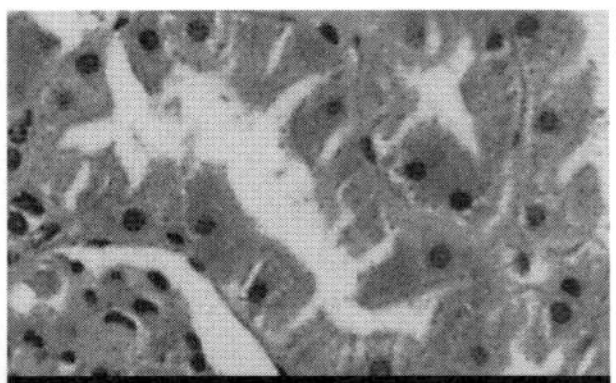

Kidney cells — F _____

Alderfly larva — G _____

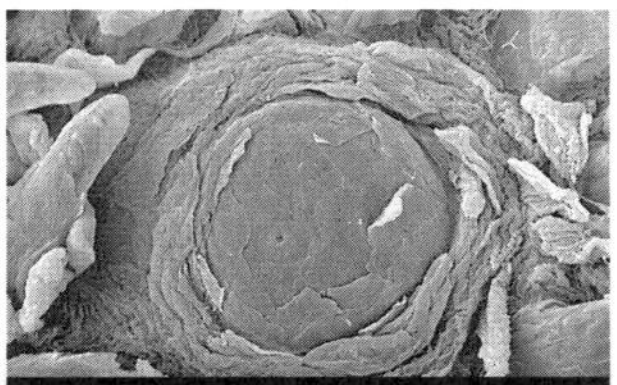

Tongue papilla — H _____

Interpreting Electron Micrographs

113

The photographs below were taken using a transmission electron microscope (TEM). They show some of the cell organelles in great detail. Remember that these photos are showing only parts of cells, not whole cells. Some of the photographs show more than one type of organelle. The questions refer to the main organelle in the center of the photo.

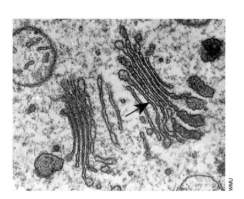

1. (a) Name this organelle (arrowed): _____

   (b) State which kind of cell(s) this organelle would be found in:

   _____

   (c) Describe the function of this organelle: _____

   _____

   _____

   (d) Label two structures that can be seen inside this organelle.

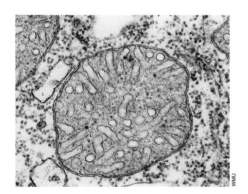

2. (a) Name this organelle (arrowed): _____

   (b) State which kind of cell(s) this organelle would be found in:

   _____

   (c) Describe the function of this organelle: _____

   _____

   _____

   _____

   _____

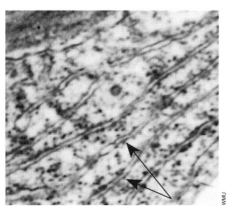

3. (a) Name the large, circular organelle: _____

   (b) State which kind of cell(s) this organelle would be found in:

   _____

   (c) Describe the function of this organelle: _____

   _____

   _____

   (d) Label two regions that can be seen inside this organelle.

4. (a) Name and label the ribbon-like organelle in this photograph (arrowed):

   _____

   (b) State which kind of cell(s) this organelle is found in:

   _____

   (c) Describe the function of these organelles: _____

   _____

   _____

   (d) Name the dark 'blobs' attached to the organelle you have labeled:

   _____

© Biozone International 2006-2007
Photocopying Prohibited

Related activities: Electron Microscopes

Laboratory Techniques

RA 1

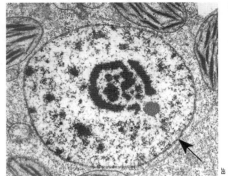

5. (a) Name this large circular structure (arrowed): _____

   (b) State which kind of cell(s) this structure would be found in:
   _____

   (c) Describe the function of this structure: _____
   _____
   _____

   (d) Label three features relating to this structure in the photograph.

6. The four dark structures shown in this photograph are called **desmosomes**. They cause the plasma membranes of neighboring cells to stick together. Without desmosomes, animal cells would not combine together to form tissues.

   (a) Describe the functions of the plasma membrane:
   _____
   _____
   _____

   (b) Label the plasma membrane and the four desmosomes in the photograph.

7. In the space below, draw a simple, labeled diagram of a **generalized cell** to show the relative size and location of these six structures and organelles (simple outlines of the organelles will do):

Identifying Cell Structures

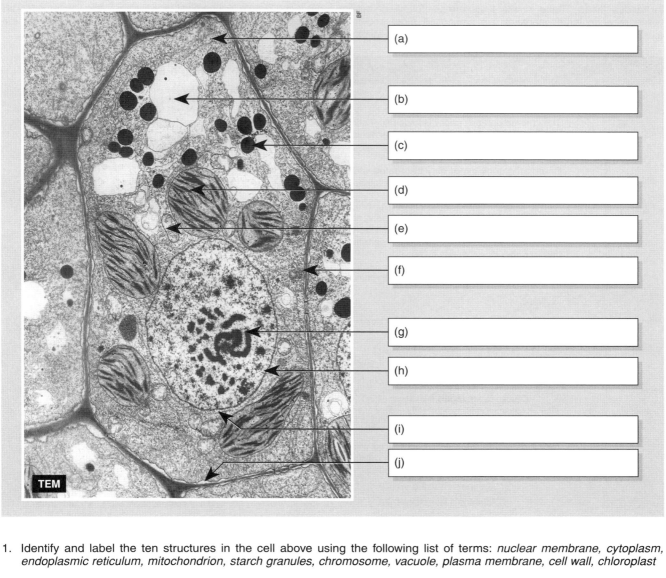

1. Identify and label the ten structures in the cell above using the following list of terms: *nuclear membrane, cytoplasm, endoplasmic reticulum, mitochondrion, starch granules, chromosome, vacuole, plasma membrane, cell wall, chloroplast*

2. State how many cells, or parts of cells, are visible in the electron micrograph above: _____

3. Identify the **type** of cell illustrated above (bacterial cell, plant cell, or animal cell). Explain your answer:
   _____
   _____

4. (a) Explain where cytoplasm is found in the cell: _____
   _____

   (b) Describe what cytoplasm is made up of: _____
   _____

5. Describe two structures, pictured in the cell above, that are associated with storage:

   (a) _____
   _____

   (b) _____
   _____

Biochemical Tests

Biochemical tests are used to detect the presence of nutrients such as lipids, proteins, and carbohydrates (sugar and starch) in various foods. These simple tests are useful for detecting nutrients when large quantities are present. A more accurate technique by which to separate a mixture of compounds involves chromatography. Chromatography is used when only a small sample is available or when you wish to distinguish between nutrients. Simple biochemical food tests will show whether sugar is present, whereas chromatography will distinguish between the different types of sugars (e.g. fructose or glucose).

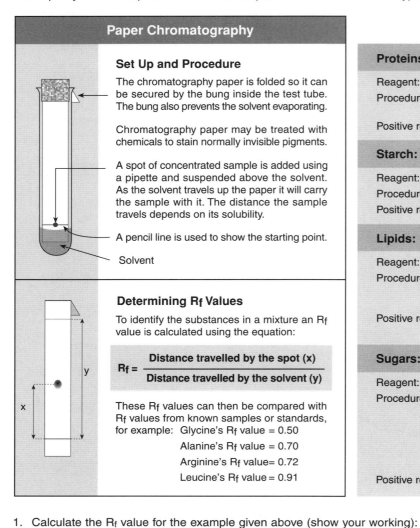

Paper Chromatography

Set Up and Procedure

The chromatography paper is folded so it can be secured by the bung inside the test tube. The bung also prevents the solvent evaporating.

Chromatography paper may be treated with chemicals to stain normally invisible pigments.

A spot of concentrated sample is added using a pipette and suspended above the solvent. As the solvent travels up the paper it will carry the sample with it. The distance the sample travels depends on its solubility.

A pencil line is used to show the starting point.

Solvent

Determining R_f Values

To identify the substances in a mixture an R_f value is calculated using the equation:

$$R_f = \frac{\text{Distance travelled by the spot (x)}}{\text{Distance travelled by the solvent (y)}}$$

These R_f values can then be compared with R_f values from known samples or standards, for example: Glycine's R_f value = 0.50
Alanine's R_f value = 0.70
Arginine's R_f value = 0.72
Leucine's R_f value = 0.91

Simple Food Tests

Proteins: The Biuret Test

Reagent: Biuret solution.
Procedure: A sample is added to biuret solution and gently heated.
Positive result: Solution turns from blue to lilac.

Starch: The Iodine Test

Reagent: Iodine.
Procedure: Iodine solution is added to the sample.
Positive result: Blue-black staining occurs.

Lipids: The Emulsion Test

Reagent: Ethanol.
Procedure: The sample is shaken with ethanol. After settling, the liquid portion is distilled and mixed with water.
Positive result: The solution turns into a cloudy-white emulsion of suspended lipid molecules.

Sugars: The Benedict's Test

Reagent: Benedict's solution.
Procedure: *Non reducing sugars*: The sample is boiled with dilute hydrochloric acid, then cooled and neutralised. A test for reducing sugars is then performed.
Reducing sugar: Benedict's solution is added, and the sample is placed in a water bath.
Positive result: Solution turns from blue to orange.

1. Calculate the R_f value for the example given above (show your working): _____

2. Explain why the R_f value of a substance is always less than 1: _____

3. Discuss when it is appropriate to use chromatography instead of a simple food test: _____

4. Predict what would happen if a sample was immersed in the chromatography solvent, instead of suspended above it: _____

5. With reference to their R_f values, rank the four amino acids (listed above) in terms of their solubility: _____

6. Outline why lipids must be mixed in ethanol before they will form an emulsion in water: _____

Differential Centrifugation

Differential centrifugation (also called cell fractionation) is a technique used to extract organelles from cells so that they can be studied. The aim is to extract undamaged intact organelles. Samples must be kept very cool so that metabolism is slowed and self digestion of the organelles is prevented. The samples must also be kept in a buffered, isotonic solution so that the organelles do not change volume and the enzymes are not denatured by changes in pH.

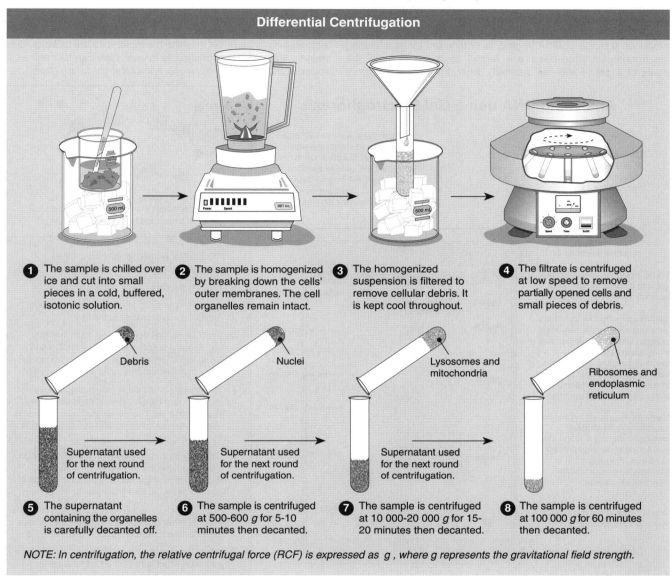

Differential Centrifugation

1. The sample is chilled over ice and cut into small pieces in a cold, buffered, isotonic solution.
2. The sample is homogenized by breaking down the cells' outer membranes. The cell organelles remain intact.
3. The homogenized suspension is filtered to remove cellular debris. It is kept cool throughout.
4. The filtrate is centrifuged at low speed to remove partially opened cells and small pieces of debris.
5. The supernatant containing the organelles is carefully decanted off.
6. The sample is centrifuged at 500-600 g for 5-10 minutes then decanted.
7. The sample is centrifuged at 10 000-20 000 g for 15-20 minutes then decanted.
8. The sample is centrifuged at 100 000 g for 60 minutes then decanted.

NOTE: In centrifugation, the relative centrifugal force (RCF) is expressed as g, where g represents the gravitational field strength.

1. Explain why it is possible to separate cell organelles using centrifugation: _____

2. Suggest why the sample is homogenized before centrifugation: _____

3. Explain why the sample must be kept in a solution that is:

   (a) Isotonic: _____

   (b) Cool: _____

   (c) Buffered: _____

4. **Density gradient centrifugation** is another method of cell fractionation. Sucrose is added to the sample, which is then centrifuged at high speed. The organelles will form layers according to their specific densities. Using the information above, label the centrifuge tube on the right with the organelles you would find in each layer.

Density gradient centrifugation

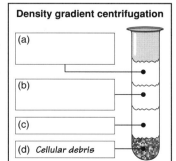

(a)
(b)
(c)
(d) Cellular debris

Gel Electrophoresis

Gel electrophoresis is a method that separates large molecules (including nucleic acids or proteins) on the basis of size, electric charge, and other physical properties. Such molecules possess a slight electric charge (see DNA below). To prepare DNA for gel electrophoresis the DNA is often cut up into smaller pieces. This is done by mixing DNA with restriction enzymes in controlled conditions for about an hour. Called **restriction digestion**, it produces a range of DNA fragments of different lengths. During electrophoresis, molecules are forced to move through the pores of a **gel** (a jelly-like material), when the electrical current is applied. Active electrodes at each end of the gel provide the driving force. The electrical current from one electrode repels the molecules while the other electrode simultaneously attracts the molecules. The frictional force of the gel resists the flow of the molecules, separating them by size. Their rate of migration through the gel depends on the strength of the electric field, size and shape of the molecules, and on the ionic strength and temperature of the buffer in which the molecules are moving. After staining, the separated molecules in each lane can be seen as a series of bands spread from one end of the gel to the other.

Analysing DNA using Gel Electrophoresis

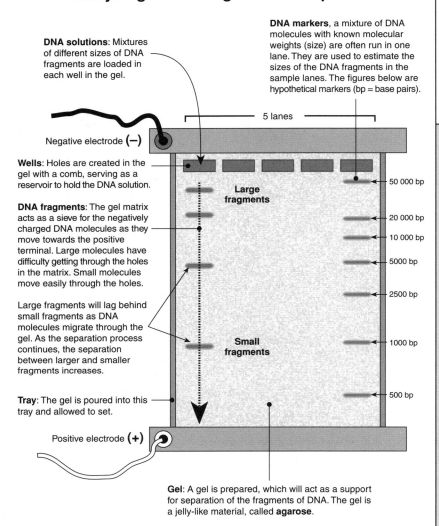

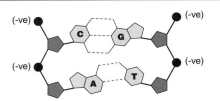

DNA is negatively charged because the phosphates (black) that form part of the backbone of a DNA molecule have a negative charge.

Steps in the process of gel electrophoresis of DNA

1. A tray is prepared to hold the gel matrix.
2. A gel comb is used to create holes in the gel. The gel comb is placed in the tray.
3. Agarose gel powder is mixed with a buffer solution (the liquid used to carry the DNA in a stable form). The solution is heated until dissolved and poured into the tray and allowed to cool.
4. The gel tray is placed in an electrophoresis chamber and the chamber is filled with buffer, covering the gel. This allows the electric current from electrodes at either end of the gel to flow through the gel.
5. DNA samples are mixed with a "loading dye" to make the DNA sample visible. The dye also contains glycerol or sucrose to make the DNA sample heavy so that it will sink to the bottom of the well.
6. A safety cover is placed over the gel, electrodes are attached to a power supply and turned on.
7. When the dye marker has moved through the gel, the current is turned off and the gel is removed from the tray.
8. DNA molecules are made visible by staining the gel with **methylene blue** or ethidium bromide (which binds to DNA and fluoresces in UV light).

1. Explain the purpose of gel electrophoresis: _____

2. Describe the two forces that control the speed at which fragments pass through the gel:

    (a) _____

    (b) _____

3. Explain why the smallest fragments travel through the gel the fastest: _____

Analyzing a DNA Sample

The nucleotide (base sequence) of a section of DNA can be determined using DNA sequencing techniques (see the modular workbook called *Microbiology and Biotechnology* for a description of this technology). The base sequence determines the amino acid sequence of the resultant protein therefore the DNA tells us what type of protein that gene encodes. This exercise reviews the areas of DNA replication, transcription, and translation using an analysis of a gel electrophoresis column. Remember that the gel pattern represents the sequence in the synthesized strand.

1. Determine the amino acid sequence of a protein from the nucleotide sequence of its DNA, with the following steps:
   (a) Determine the sequence of **synthesized DNA** in the gel
   (b) Convert it to the complementary sequence of the **sample DNA**
   (c) Complete the **mRNA** sequence
   (d) Determine the **amino acid** sequence by using a 'mRNA amino acid table' (consult a reference source).

   NOTE: The nucleotides in the gel are read from bottom to top and the sequence is written in the spaces provided from left to right (the first 4 have been done for you).

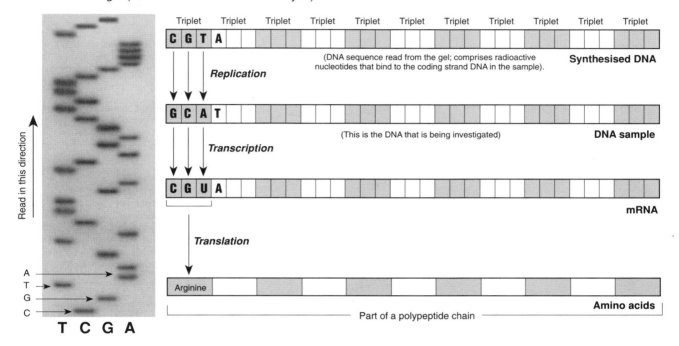

2. For each single strand DNA sequence below, write the base sequence for the **complementary DNA** strand:

   (a) DNA: T A C   T A G   C C G   C G A   T T T   A C A   A T T

       DNA: _____

   (b) DNA: T A C   G C C   T T A   A A G   G G C   C G A   A T C

       DNA: _____

   (c) Identify the cell process that this exercise represents: _____

3. For each single strand DNA sequence below, write the base sequence for the **mRNA** strand and the **amino acid** that it codes for (refer to a mRNA amino acid table to determine the amino acid sequence):

   (a) DNA: T A C   T A G   C C G   C G A   T T T   A C A   A T T

       mRNA: _____

       Amino
       acids: _____

   (b) DNA: T A C   G C C   T T A   A A G   G G C   C G A   A T C

       mRNA: _____

       Amino
       acids: _____

   (c) Identify the cell process that this exercise represents: _____

Techniques in Microbial Culture

Bacteria and fungi may be cultured in liquid or solid media. These comprise a base of **agar** to which is added the nutrients required for microbial growth. Agar is a gelatinous colloidal extract of red algae, and can be used in solid or liquid form. It is used because of its two unique physical properties. Firstly, it melts at 100°C and remains liquid until cooled to 40°C, at which point it gels. Secondly, few microbes are capable of digesting agar so the medium is not used up during culture. The addition of microbes to an agar plate, or to liquid agar, is called **inoculation** and must be carried out under aseptic conditions. **Aseptic techniques** involve the **sterilization** of equipment and culture media to prevent cross contamination by unwanted microbes. Sterilization is a process by which all organisms and spores are destroyed, either by heat or by chemicals.

Conditions for the Culture of Bacteria and Fungi

Fungi

Temperature: Most fungi have an optimum temperature for growth of 25°C, but most are adapted to survive between 5 and 35°C.

pH: Fungi prefer a neutral (pH 7) growing environment, although most species can tolerate slightly acidic conditions.

Nutrients: Fungi require a source of carbon and nitrogen to produce protein. They also require trace elements such as potassium, phosphorus and magnesium. Growth factors can be added to increase the rate of fungal growth.

Water potential: Fungi are 85-90% water by mass. Water is constantly lost from the hyphae via evaporation and must be replaced through absorption from the media. To aid water uptake, media have a water potential that is less negative than that of the fungal tissue.

Gaseous environment: The majority of fungi are aerobic and very few species can tolerate anaerobic conditions. This is why fungi always grow on the surface of a culture medium, not inside it.

Bacteria

Temperature: Most bacteria cultured in the school laboratory are classified as **mesophiles**. Mesophiles prefer temperatures between 20 and 40°C.

pH: Most bacteria grow optimally in media with a pH between 6 and 8. Very few bacteria can grow in acidic conditions.

Nutrients: Bacteria need a source of carbon, nitrogen and mineral salts as raw ingredients for cellular growth.

Water potential: All bacteria require water for growth. To prevent cell lysis or dehydration, the water potential of the medium must be such that net water fluxes into and out of the bacterial cell are minimized.

Gaseous environment: Aerobic bacteria will grow only in oxygenated environments, whereas obligate anaerobes (e.g. *Clostridium*) do not tolerate oxygen. Facultative anaerobes grow under aerobic conditions, but are able to metabolize anaerobically when oxygen is unavailable. All bacterial cultures benefit from a low concentration of carbon dioxide.

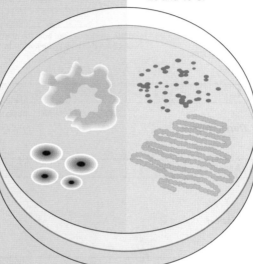

Inoculating Solid Media

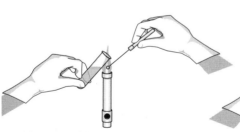

1. Hold the inoculating loop in the flame until it glows red hot. Remove the lid from the culture broth and pass the neck of the bottle through the flame.

2. Dip the cool inoculating loop into the broth. Flame the neck of the bottle again and replace the lid.

3. Raise the lid of the plate just enough to allow the loop to streak the plate. Streak the surface of the media. Seal the plate with tape and incubate upside down.

1. Explain why inoculated plates must be stored upside down in an incubator: _____

2. Outline the correct procedure for the disposal of microbial plates and cultures: _____

3. Suggest a general method by which you could separate microorganisms through culturing: _____

RA 1   **Related activities**: Strain Isolation, Serial Dilution

Strain Isolation

In nature, bacteria exist as mixed populations. However, in order to study them in the laboratory they must exist as pure cultures (i.e. cultures in which all organisms are descendants of the same organism or clones). The most common way of separating bacterial cells on the agar surface is the **streak plate method**. This provides a simple and rapid method of diluting the sample by mechanical means. As the loop is streaked across the agar surface, more and more bacteria are rubbed off until individual separated organisms are deposited on the agar. After incubation, the area at the beginning of the streak pattern will show **confluent** growth (growth as a continuous sheet), while the area near the end of the pattern should show discrete colonies. Isolated colonies can then be removed from the streak plate using aseptic techniques, and transferred to new sterile medium. After incubation, all organisms in the new culture will be descendants of the same organism (i.e. a pure culture).

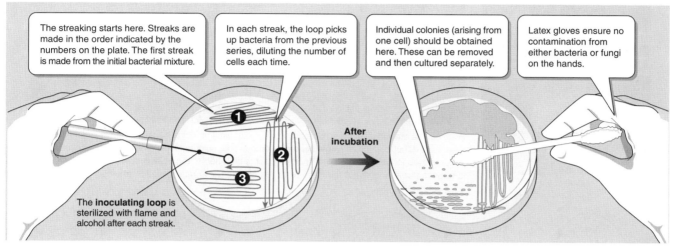

When approximately 10 to 100 million bacterial cells are present, colonies become visible. Note the well-isolated colonies in the photo above. A single colony may be removed for further investigation.

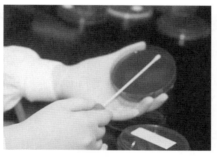

A swab containing a single strain of bacteria is used to inoculate additional nutrient plates to produce pure cultures of bacteria (clones).

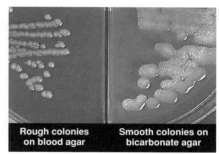

To test purity, a sample of a culture can be grown on a selective medium that promotes the growth of a single species. The photo above shows a positive encapsulation test for *Bacillus anthracis*.

1. Explain the basis by which bacteria are isolated using streak plating: _____

2. Discuss the basic principles of aseptic technique, outlining why each procedure is necessary: _____

3. Comment on the importance of aseptic (sterile) technique in streak plating: _____

4. State how many bacterial cells must be present on the plate before the colony becomes visible to the naked eye: _____

5. Outline when it might be necessary to use **selective media** to culture bacteria: _____

Serial Dilution

The growth of microorganisms in culture can be measured in a number of ways. Some indirect methods measure culture dry weight or turbidity, both of which are often directly proportional to cell density. More commonly used are methods that directly or indirectly count the number of cells in a culture. Microbial populations are often very large, so most counting methods rely on counting a very small sample of the culture. A commonly used indirect method is serial dilution followed by plate counts (below). If care is taken with the serial dilution, this method can provide a relatively accurate estimate of culture density.

Measuring Microbial Growth Using Serial Dilution

Serial dilution can be performed at different stages during the culture growth. By making a series of dilutions and then counting the colonies that arise after plating, the density of the original inoculum (starting culture) can be calculated. Colonies should be well separated and the number of colonies counted should ideally be neither too small nor too large (about 15-30 is good).

CALCULATION: No. of colonies on plate X reciprocal of sample dilution = no. of bacteria per cm^3.

EXAMPLE: 28 colonies on a plate of 1/1000 dilution, then the original culture contained:
$28 \times 1000 = 28 \times 10^3$ cm^{-3} bacterial cells

Plate counts are widely used in microbiology. It is a useful technique because only the viable colonies are counted, but it requires some incubation time before colonies form. For quality control purposes in some food industries where the food product is perishable (e.g. milk processing) this time delay is unacceptable and direct methods (e.g. cell counts using oil immersion microscopy) are used.

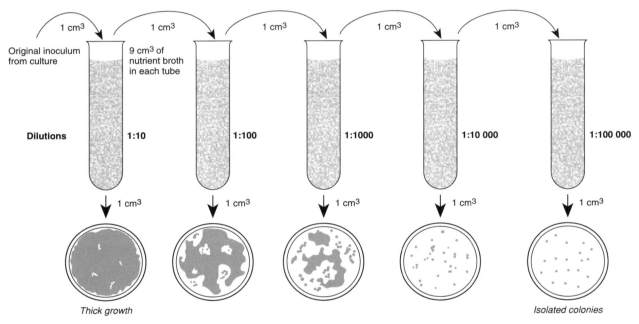

1. In the example of serial dilution above, use the equation provided to calculate the cell concentration in the original culture:
   _____

2. (a) Explain the term **viable count**: _____
   _____

   (b) Explain why dilution plating is a useful technique for obtaining a viable count: _____
   _____
   _____

   (c) Investigate an alternative technique, such as turbidimetry and identify how the technique differs from dilution plating:
   _____
   _____
   _____
   _____

Plant Tissue Culture

Plant tissue culture, or **micropropagation**, is a method used to reproduce plants under sterile conditions to produce **clones**. It is used widely for the rapid multiplication of commercially important plant species with superior genotypes, as well as in the recovery programs for endangered plant species. Micropropagation relies on the fact that differentiated plant cells have the potential to give rise to all the cells of an adult plant. This property, called **totipotency**, means that single cells (**protoplasts**) and small pieces of plant tissue (**explants**) can be used to generate a new plant on culture media. Micropropagation has considerable advantages over traditional methods of plant propagation (such as cutting or grafting), but it is very labor intensive and the optimal conditions for growth and development must be determined and reproduced in the laboratory.

Basic Steps in Tissue Culture

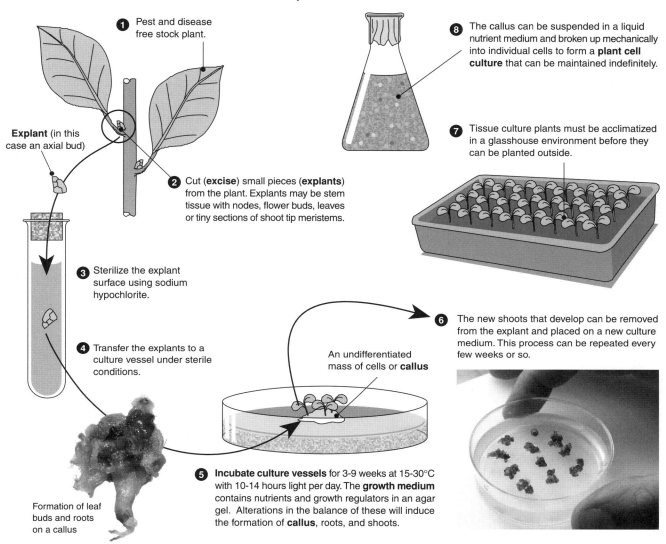

1. Pest and disease free stock plant.
2. Cut (**excise**) small pieces (**explants**) from the plant. Explants may be stem tissue with nodes, flower buds, leaves or tiny sections of shoot tip meristems.
3. Sterilize the explant surface using sodium hypochlorite.
4. Transfer the explants to a culture vessel under sterile conditions.
5. **Incubate culture vessels** for 3-9 weeks at 15-30°C with 10-14 hours light per day. The **growth medium** contains nutrients and growth regulators in an agar gel. Alterations in the balance of these will induce the formation of **callus**, roots, and shoots.
6. The new shoots that develop can be removed from the explant and placed on a new culture medium. This process can be repeated every few weeks or so.
7. Tissue culture plants must be acclimatized in a glasshouse environment before they can be planted outside.
8. The callus can be suspended in a liquid nutrient medium and broken up mechanically into individual cells to form a **plant cell culture** that can be maintained indefinitely.

1. Explain the importance of totipotency to micropropagation: _____

2. Explain how root and shoot formation is induced in a callus: _____

3. Outline some the **advantages** of micropropagation over traditional methods of plant propagation: _____

Index

Analyzing results 37-38
Animals, features of 88-90, 103
Animals, sampling populations of 75-76
Analysis of Variance (ANOVA) 49-51
Archaea 84, 86
Aseptic technique 120
Assumptions 15, 17

Bacteria
- classification of 84
- culture of 120
- features of 86, 102

Bar graphs 26-27
Belt transects 64, 73
Benedict's test 116
Binomial nomenclature 91-93
Biochemical tests 116
Biological drawings 107-108
Biuret test 116

Cell fractionation 117
Chi-squared test 52-55
Chromatography 116
Citation, of references 61-62
Cloning, of plants 123
Cladistics 85
Classification
- cladistic 85
- domains 84
- five kingdom 84, 96
- keys 93-95
- of hedgehog 92
- of life 97
- schemes 85
- system of 91-92

Comparing more than two groups 49
Compound microscopes 109, 112
Confidence intervals 41
Continuous data 14
Controlled variables 17

Data 14, 39
- distribution of 39
- measuring spread in 39
- presentation 22
- statistical analysis of 38
- transformation 23

Dataloggers 67
Dependent variable 17
Descriptive statistics 39-42
Differential centrifugation 117
Discontinuous data 14
Discussion writing 59
Dissecting microscopes 109
Distribution of data 39
DNA, analyzing a sample of 119
Drawings, biological 107-108

Electron micrographs, interpretation of 113-115
Electron microscopes 111-112
Emulsion test 116
Environmental gradient 73
Eubacteria 84, 86
Eukarya (Eukaryota) 84
Experimental design 19

Field study design 65-66
Frequency table 23
Fungal culture 117
Fungi, features of 86, 96, 105

Gel electrophoresis 118-119
Graphs
- interpretation of 35
- presentation of 22
- types of 25-33

Growth, of microbes 122

Histograms 26, 28
Hypotheses
- forming 15
- types of 15

Independent variable 17
Indirect sampling 79-80
Inoculation, of plates 120-121
Investigations flow chart 37
Iodine test 116

Keys, classification 93-94
Kite graphs 26, 30, 74
Kingdoms, features of 84, 86-90, 96

Leaf litter population, sampling 71
Lincoln index 77
Line graphs 25, 32-33
Line transects 64, 73
Linear regression 43

Mark and recapture sampling 64, 77
Mean (average), of data 39
Media, in microbial growth 120
Median, of data 39
Meters, for monitoring physical factors 67
Methods, writing 57
Microbial cultures 120
Microbial groups, features of 86
Micropropagation 123
Microscopes
- dissecting 109
- drawings using 107-108
- electron 111-112
- optical 109, 112

Mode, of data 39

Non-linear regression 45
Notation, scientific 10
Null hypothesis 11

Observations 11, 15
Optical microscopes 109, 112

Percentage, calculating 23
Phylogeny 85
Physical factors, monitoring 67
Pie graphs 26, 29
Planning an investigation 17-18
Plants, features of 87, 101, 105
Plant tissue culture 123
Plate counts, of bacteria 122
Point sampling 64, 73
Populations
- density calculation of 69, 71
- sampling 64, 75

Predictions 11, 15
Prokaryotes, features of 86, 101-102
Protists, features of 86, 101-102

Quadrat sampling 64, 69-70
Quadrat-based estimates 70
Qualitative data 14, 39
Quantitative data 14, 39

Radio-tracking 76, 81-82
Random number table 72
Range, of data 39
Rate, calculating 23
Reciprocal, calculating 23
Recording results 21
Referencing 61-62
Regression 43, 45
Relative value, calculating 23
Reliability, of the mean 41
Replication, in experiments 20

Report writing 56, 60
Research project guide 13
Resolution 109
- of different microscopes 112
Restriction digestion 118
Results tables 21-22
Results, writing 58
R_f value 116

Sample size 66
Sampling
- animals 75-76
- populations 64, 75
- using quadrats 64, 69-70
- using radio-tracking 76, 81-82
- using transects 64, 73

Scanning electron microscope (SEM) 111-112
Scatter plots 25, 31
Scientific method 11
Scientific notation 10
Serial dilution 122
Species abundance scales 70
Species, classification of 91
Stains, in microscopy 110
Standard deviation 39, 41
Standard error 39, 41
Statistics
- descriptive 39-42
- flow chart 38

Strain isolation, of bacteria 121
Streak plating 121
Student's t test 46-48

t test 46-48
Tables, presentation 22
Taxonomic groups, features of 86-90, 101-105
Taxonomy 85
Techniques in microbial culture 120
Terms, scientific 10
Tissue culture 123
Topic selection for project 14
Total, calculating 23
Transect sampling 64, 73
Transmission electron microscope 111-112
Transmitter design 82

Variables
- defining 14
- identifying 17